Asynchronous Sequential Machine Design and Analysis

A Comprehensive Development of the Design and Analysis of Clock–Independent State Machines and Systems

Asynchronous Sequential Machine Design and Analysis: A Comprehensive Development of the Design and Analysis of Clock-Independent State Machines and Systems
Richard F. Tinder

ISBN: 978-3-031-79787-3   paperback

ISBN: 978-3-031-79788-0   ebook

DOI: 10.1007/978-3-031-79788-0

A Publication in the Springer series

SYNTHESIS LECTURES ON DIGITAL CIRCUITS AND SYSTEMS #18

Lecture #18

Series Editor: Mitchell Thornton, Southern Methodist University

Series ISSN
ISSN 1932-3166   print
ISSN 1932-3174   electronic

Asynchronous Sequential Machine Design and Analysis

A Comprehensive Development of the Design and Analysis of Clock–Independent State Machines and Systems

Richard F. Tinder
Professor Emeritus of Electrical Engineering and Computer Science,
Washington State University

SYNTHESIS LECTURES ON DIGITAL CIRCUITS AND SYSTEMS #18

ABSTRACT

Asynchronous Sequential Machine Design and Analysis provides a lucid, in-depth treatment of asynchronous state machine design and analysis presented in two parts: Part I on the background fundamentals related to asynchronous sequential logic circuits generally, and Part II on self-timed systems, high-performance asynchronous programmable sequencers, and arbiters.

Part I provides a detailed review of the background fundamentals for the design and analysis of asynchronous finite state machines (FSMs). Included are the basic models, use of fully documented state diagrams, and the design and characteristics of basic memory cells and Muller C-elements. Simple FSMs using C-elements illustrate the design process. The detection and elimination of timing defects in asynchronous FSMs are covered in detail. This is followed by the array algebraic approach to the design of single-transition-time machines and use of CAD software for that purpose, one-hot asynchronous FSMs, and pulse mode FSMs. Part I concludes with the analysis procedures for asynchronous state machines.

Part II is concerned mainly with self-timed systems, programmable sequencers, and arbiters. It begins with a detailed treatment of externally asynchronous/internally clocked (or pausable) systems that are delay-insensitive and metastability-hardened. This is followed by defect-free cascadable asynchronous sequencers, and defect-free one-hot asynchronous programmable sequencers—their characteristics, design, and applications. Part II concludes with arbiter modules of various types, those with and without metastability protection, together with applications.

Presented in the appendices are brief reviews covering mixed-logic gate symbology, Boolean algebra, and entered-variable K-map minimization. End-of-chapter problems and a glossary of terms, expressions, and abbreviations contribute to the reader's learning experience. Five productivity tools are made available specifically for use with this text and briefly discussed in this front matter.

KEYWORDS

asynchronous, sequential, sequencers, logic, machines, digital, self-timed, arbiters

Contents

Preface

This text emphasizes the design and analysis of a variety of asynchronous sequential machines and high-performance asynchronous programmable sequencers presented in a hands-on and in-depth manner. As background to this, we devote considerable effort in developing the basic models, the use of fully documented state diagrams, and the associated rules and algorithms needed to carry out rather complex designs and analyses. The analysis and elimination of the timing defects, those exclusively owned by asynchronous sequential machines, are an important part of the background fundamentals developed in this text. We also emphasize the use of Muller C-elements, which operate outside of the fundamental mode, and use them in various different design approaches.

Inclusion of asynchronous systems in modern computer, microprocessor, and application-specific integrated circuit chip designs have become a reality. With the advent of picosecond complementary metal-oxide semiconductor (CMOS) technology, reliable high-speed asynchronous performance is now a reality. We combine the use of asynchronous and locally clocked controller systems, thereby offering an attractive alternative to conventional synchronous systems. This is accomplished in such a manner as to eliminate clock skew and improve reliability while operating with a significant reduction in the power-delay product. Moreover, we develop the use of cascadable asynchronous programmable sequencers that can be made reprogrammable during operation, permitting instantaneous changes between radically different asynchronous state machines on a time-shared basis, all timing-defect-free. Included is the development of a variety of arbiters that can be used with both Huffman and Muller controller/data-path frameworks.

The contents of this text are based on the author's lecture notes used in a graduate course taught over many years at Washington State University to graduate students and second-semester seniors in electrical and computer engineering. The text is designed to be used as a one-semester or two-quarter course in the subject matter. It also serves as a valuable source of information for practicing engineers and computer scientists in related fields. Although the text provides the necessary background in asynchronous sequential machine design and analysis, the readership is expected to have had a beginning course in logic design and some knowledge of basic Boolean algebra and K-map minimization. Every effort is made to take this complex subject matter and present it in a manner that can be understood by a readership consisting of individuals of varying backgrounds.

An instructor of a course in this subject matter is provided ample opportunity to limit, alter, or expound on any part of the text material as needed to satisfy the course description and needs of the students. In this regard, the use of Very High Speed IC Hardware Description Language (VHDL) or Verilog for purposes of circuit representation and simulation is encouraged. However, space limitation prohibits their development in this text. End-of-chapter problems and a glossary of terms, expressions, abbreviations, and symbols at the end of the text aid in the learning process.

A word to the readership: Due to space limitations, the information and developments found in this text are presented in a thoughtful but succinct manner requiring careful reading. To assist in the learning process, each subject is accompanied by the appropriate state diagrams, logic schematics, simulations, and Boolean functions as needed for clarity and emphasis. Numerous tables also contribute in this regard.

Acknowledgments

The author is deeply indebted to the many senior and graduate students for their countless suggestions, contributions, comments, and arguments given during the many years the content of this text were used in the conduct of the author's graduate level course at Washington State University. As most professors know, students can and do provide the most candid and critical assessments of material presented in a given course. Such involvement has definitely enhanced the pedagogy of this book's subject matter. In this regard, the graduate student research contributions of Anders Boen, Shawn Galiher, Hani Khalil, Rick Klaus, Thomas Kovacs, Gerald Murphy, and William VanScheik are gratefully acknowledged. The author also acknowledges the insight and suggestions of Prof. Mark Manwaring, formerly professor of Computer Engineering at Brigham Young University and now chairman of Computer Science at the University of Idaho. Dr. Manwaring, as conference chair, provided the author with the opportunity to present some of the text material at the First Annual David C. Evans Computer Engineering Conference held in Salt Lake City, Utah, on June 3–4, 2002. The author would also like to acknowledge the useful discussions with Prof. Chris J. Myers of the University of Utah. A debt of gratitude must be attributed to Prof. George LaRue of Washington State University for his help with the SPICE study involving possible metastable conditions in Muller C-elements.

The encouragement and helpful suggestions offered by Joel Claypool, publisher of Morgan & Claypool Publishers, is also gratefully acknowledged. Finally and most importantly, the author wishes to acknowledge the support, care, and understanding of his loving wife, best friend, and confidante, Gloria.

Instructional Support Software

Students and faculty alike will find these five productivity tools highly useful, if not essential, in solving many end-of-chapter problems that appear in this text. Complete instructions accompany each software program as a Readme_<*software name*>.doc WORD file. All of the software described below, except the EXL-Sim simulator, require the use of a text editor and can be downloaded from http://www.exlsim.com cost free. The student version of EXL-Sim used in this text can be purchased from this website at nominal cost.

(1) *EXL-Sim logic simulator*. EXL-Sim is a full-feature, interactive, schematic-capture, and simulation program that is ideally suited for use with this text at either the entry or advanced level of logic design. It is the student version of a more powerful professional-level program. Its many features include a complete library of mixed logic gate symbols, Muller C-elements with or without CLEAR and user-defined delay and buffer symbols; drag-and-drop capability, part rotation and mirroring, circuit and part duplication, rubber banding, mixed-logic or positive logic simulations (positive logic waveforms mimic those of voltage waveforms); multiple levels of macro generation; individual input, gate, and global delay assignments, a wireless connection feature that eliminates the need to use wire connections and minimizes error; labeling, editing, sizing, and zooming of schematics; waveform interactive editing, zooming, scrolling, animation, and stepping; library and project management; a variety of export and printout capabilities; and a host of other features including preferences for default settings. Check http://www.exlsim.com for updates and announcements.

Note on the schematic captures and simulations of EXL-Sim used in this text: Many of the logic schematics and labels have been produced by using a graphics program so as to bring the symbology in agreement with that used in other parts of the text. For example, EXL-Sim cannot produce subscripts but the graphics program can, and subscripts are used in the text. However, there are many other schematics that have been imported into WORD as produced by EXL-Sim. We normally do so to distinguish between the two methods. The simulations are precisely that produced by EXL-Sim but the labels are usually done by the graphics program before importing them into WORD.

For some simulations, there are additional labels needed for clarification that had to be added by the graphics program and superimposed on the waveforms.

Note on the use of gate path delays: For simplicity, we have used the same path delays for all gates throughout most of the text. However, we could have modeled them close to each physical gate or used random delays, although these added features did not seem to add any significant advantages in conveying the information to the reader.

(2) *BOOZER logic minimizer*. BOOZER is a software minimization tool that is recommended for use with this text. It accepts entered variable or canonical (1's and 0's) data from K-maps or truth tables, with or without "don't cares," and returns an optimal or near-optimal single or multioutput solution. It can handle up to 12 Boolean function outputs and as many inputs when used on modern computers.

(3) *ESPRESSO II logic minimizer*. ESPRESSO II is another software minimization tool that is in wide use in schools and industry. It supports advanced heuristic algorithms for minimization of two-level, multioutput Boolean functions, but accepts canonical data only. It is also readily available from the University of California, Berkeley, 1986 VLSI Tools Distribution.

(4) *ADAM—Advanced CAD design software*. Automated Design of Asynchronous Machines (ADAM) is a very powerful productivity tool that permits the automated design of complex asynchronous single-transition-time state machines, all free of timing defects. The input files are state tables for the desired state machines. The output files can be given in the Berkeley format appropriate for directly programming PLAs. ADAM also allows the designer to design synchronous state machines, timing-defect-free. For asynchronous FSM designs, the options include the lumped path delay model and the nested element model applicable to Muller C-elements or basic SR cells. ADAM can also be used for the D flip-flop design of synchronous FSM designs.

(5) *A-OPS—Advanced CAD design software*. Asynchronous One-hot Programmable Sequencers (A-OPS) is another very powerful productivity tool that permits the design of asynchronous and synchronous state machines by using a one-hot programmable sequencer kernel. This software generates a PLA or PAL output file (in Berkeley format) or the Very High Speed IC Hardware Description Language (VHDL) code for the automated timing-defect-free designs of the following: (a) any one-hot programmable sequencer kernel up to 10 states used to drive any one-hot FSM up to 10 states; (b) the one-hot design of multiple input asynchronous or synchronous state

machines driven by either PLDs or RAM. The input file is that of a next-state table, for the desired state machine, taken from a state diagram or state table. This software can be used to design systems with the capability of instantly switching between several radically different controllers on a time-shared basis and all defect-free. An optional essential hazard analysis is provided for each one-hot design.

PART I

Background Fundaments for Design and Analysis of Asynchronous State Machines

CHAPTER 1

Introduction and Background

1.1 FEATURES OF AND NEED FOR ASYNCHRONOUS FINITE STATE MACHINES

Although all sequential machines have certain characteristics in common, there are also specific features characterizing asynchronous finite state machines (FSMs).

- Memory in the absence of clock-driven flip-flops.
- The appearance of combinational logic circuits with feedback.

There are, of course, other characteristics exclusively attributed to asynchronous FSMs, and these will become evident as we consider in detail the intricacies of clock-independent state machines.

It is natural for us to believe that data processing in the passage through a sequential system is best accomplished by a system clock, an enabling or sampling function. This is what happens in synchronous (clock-driven) sequential systems that have led to countless remarkable accomplishments over time, many of which we are familiar with. These days, however, clock distribution problems have emerged to greatly limit some applications of large, complex, and fast synchronous systems. This has prompted inclusion of asynchronous components into synchronous system designs. The reason for this is attributed to a number of factors:

- The speed requirements of a sequential system design may exceed the capability of a clock-controlled approach. In some specific cases, a properly designed asynchronous FSM will operate faster and be more appropriate than its synchronous counterpart.
- Many clock distribution (clock skew) problems can be eliminated by the inclusion of asynchronous components in synchronous systems. Clearly, a purely asynchronous sequential machine cannot have clock skew. Remember that clock skew problems can lead to system failure by causing some components in a synchronous system to become out of synch with other parts.
- The absence of flip-flops and oscillator circuits in an asynchronous sequential system design can reduce the real estate required on an integrated circuit chip for some applications.

- Just as there are some designs that lend themselves to a synchronous design, there are others that are best designed by using an asynchronous machine approach. Because of this, we are seeing the increased embedded usage of asynchronous sequential components into synchronous system designs.

Because of these features, designers are becoming more familiar with asynchronous sequential machine design and analysis, a fact that is certain to play an important role in future super high-speed microprocessors and computers designs.

1.2   FUNDAMENTAL MODE OF OPERATION AND LUMPED PATH DELAY MODELS

Data transport through any FSM is not instantaneous; it takes time to "settle in," that is, to stabilize. This fact leads to what is now referred to as the *fundamental mode* of operation, where input *set-up time* and *hold time* requirements must be met (see Glossary). It states as follows:

Fundamental Mode of FSM Operation

Operation in the fundamental mode requires that no external FSM input may change until *all* internal signals have stabilized.

This means that only one input is permitted to change at any given time, and that each change be minimally separated in time from the next. With few exceptions, all asynchronous FSMs must adhere to this requirement. The exceptions include FSMs designed with memory elements that operate outside of the fundamental mode as will be discussed later in this and other chapters.

The *lumped path delay* (LPD) model for an asynchronous Mealy FSM is presented in Figure 1.1. It consists of a *next state* (NS) forming logic section, an LPD memory stage, and an *output (OP) forming logic* section. Connecting these three sections are the *input* (IP) lines, NS lines, and *present state* (PS) feedback lines. This is called a *Mealy model* (after G. H. Mealy), because the IP lines can directly affect both the NS and OP forming logic sections. Thus, the Mealy model is the most general model for an FSM. If the IP lines to the OP section are dropped, the model is called a *Moore model*, in honor of E. F. Moore.

In the LPD model, the delays in the next state forming logic are represented as fictitious LPD memory elements where each memory element separates an NS variable Y_i from a PS variable y_i, as shown in Figure 1.1. In this sense, the NS forming logic is considered ideal, that is, void of delays with the delays lumped as *fictitious memory elements* in the NS/PS lines. Because operation

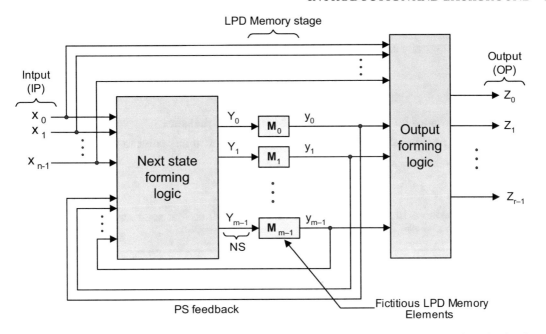

FIGURE 1.1: The generalized LPD model for an asynchronous Mealy FSM operated in the fundamental mode with fictitious LPD memory elements.

in the fundamental mode requires that no external input can be applied until any previous input has settled in and the system has stabilized, the memory in an LPD FSM is preserved. Treating the LPD model in this manner greatly simplifies the design and analysis of fundamental mode FSMs.

1.3   STABILITY CRITERIA AND THE EXCITATION TABLE FOR LPD MODELS

An inspection of Figure 1.1 indicates that the functional relationships between the stages can be represented in a simple form as

$$Y = f(\text{IP}, \text{PS}) \qquad \text{and} \qquad Z = f'(\text{IP}, \text{PS})$$

where it is understood that each LPD memory element M_i represents a fictitious delay $M_i = \Delta t_i$. It is this fictitious delay that is the cornerstone of the LPD model that, in turn, leads to the important *stability criteria* for asynchronous FSMs operated in the fundamental mode:

Stability Criteria

(a) If the PS is logically equal to the NS in a given state at some point in time, then

$$Y_i(t) = y_i(t) \qquad \text{(for all } i) \tag{1.1}$$

and the asynchronous fundamental mode FSM is *stable* in that state.

(b) If the PS is not logically equal to the NS in a given state at *any* point in time, then

$$Y_i(t) \neq y_i(t) \qquad \text{(for any } i) \tag{1.2}$$

and the asynchronous fundamental mode FSM is *unstable* in that state and must transition to another state.

Thus, the presence of an LPD memory element in each feedback loop together with operation in the fundamental mode ensures that each PS is preserved and ready for an NS transition.

Equations (1.1) and (1.2) can be represented in tabular form. When this is done, the results are the *excitation tables* shown in Figure 1.2a and 1.2b. Here, $y_t = Y_t$ signifies a stable state condition, whereas $y_t \neq Y_t$ represents an unstable state condition. The notation $y_t \rightarrow y_{t+1}$ represents a transition from the PS to the NS, implying that $y_{t+1} = Y_t$ for the NS. Readers who are familiar with synchronous FSM design will note the similarity between the excitation table for the LPD model and that for a D flip-flop. The LPD excitation table in Figure 1.2b will prove essential to the de-

Y_t	$y_t \rightarrow y_{t+1}$	
0	0 → 0	Stable
0	1 → 0	Unstable
1	0 → 1	Unstable
1	1 → 1	Stable

(a)

PS State variable change		NS variable
$y_t \rightarrow y_{t+1}$		Y_t
0 → 0		0
Set 0 → 1		1
1 → 0		0
Set Hold 1 → 1		1

(b)

FIGURE 1.2: (a) Excitation table for the LPD model as derived from Eqs. (1.1) and (1.2). (b) The excitation table of (a) arranged in the form familiar for D flip-flops now to be used for the LPD model.

sign of many asynchronous FSMs operated in the fundamental mode. This will be accomplished by combining this excitation table with a state diagram representing the sequential behavior of a FSM, the result being entered into *entered variable* (EV) *Karnaugh maps* (K-maps) and cover extracted for the design. This will all be amply discussed in this and later chapters together with the use of computer-aided design methods.

1.4 NESTED SET–RESET ELEMENT MODELS FOR ASYNCHRONOUS SEQUENTIAL MACHINES

Shown in Figure 1.3 is the generalized *nested set–reset memory element model* for asynchronous Mealy FSMs. Here, each fictitious memory element with a Y_i NS input in Figure 1.1 has been replaced by memory elements with *Set–Reset* (SR) NS inputs. The memory cells depicted in Figure 1.3 are either SR basic cells that operate in the fundamental mode, or SR *Muller C-elements* that operate outside of the fundamental mode. Although these two types of memory elements have some similar characteristics, there are important differences. For example, the basic SR cell is a simple gate-oriented FSM with one feedback line to its NS forming logic. In contrast, the SR Muller C-element is not

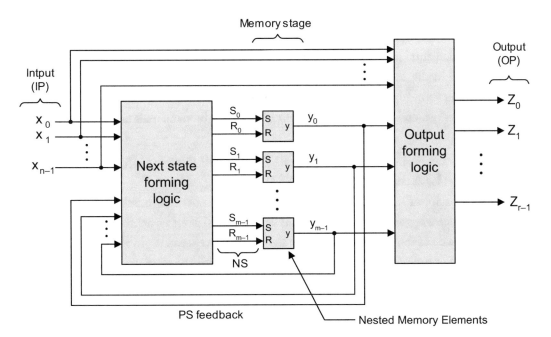

FIGURE 1.3: Nested memory element model for an asynchronous Mealy FSM showing a memory stage consisting of either basic SR cells that operate in the fundamental mode or SR C-elements that operate outside of the fundamental mode.

gate-oriented and has no feedback to its NS forming logic—it has only a weak feedback feature in its output. Use of either basic cells or C-elements in the model of Figure 1.3 may be viewed as embedded or nested FSMs within an FSM. A detailed examination of both types of memory elements is left for later discussions in this chapter. Note that if the IP lines to the OP section are dropped, the model in Figure 1.3 would be called a *nested memory element Moore model*.

Clearly, the previous discussion indicates that the memory stage can take on a different character depending on the input requirements and the type of FSM to be designed. There are four types of memory elements that can be used in the design of asynchronous FSMs:

Four Types of Memory Elements Used in Asynchronous FSM Design
(1) Fictitious LPD memory elements
(2) SR basic cells
(3) SR C-elements
(4) Toggle modules

We have touched on the first three memory elements: fictitious LPD memory elements, SR basic cells, and SR C-elements. The first two memory element types (1 and 2) are used in the design of asynchronous FSMs that must be operated in the fundamental mode. Muller C-elements operate outside of the fundamental mode, whereas toggle modules are used in the design of asynchronous FSMs that operate in the *pulse mode* discussed in Chapter 6. The use of these memory elements will be discussed, in turn, together with the appropriate design considerations. Before we do this, however, further background discussions of sequential machine fundamentals are necessary.

1.5   FULLY DOCUMENTED STATE DIAGRAM—SUM RULE AND MUTUALLY EXCLUSIVE REQUIREMENT

The sequential behavior of any FSM (asynchronous or synchronous) is revealed most effectively by a *fully documented state diagram*. Once the FSM has been declared as an asynchronous state machine operating in or outside of the fundamental mode, the design process can begin with either the LPD model (Figure 1.1) or a nested element model (Figure 1.3), with the state diagram, and with the appropriate excitation table. The excitation table in Figure 1.2 applies only to the LPD model. The excitation table for the nested SR element models is yet to be determined.

Shown in Figure 1.4 is a segment of a fully documented state diagram that represents the partial sequential behavior of an asynchronous FSM. There are three states shown, each depicted as an oval-shaped symbol. The branching paths, documented branching conditions, PS state code assignments, and the *conditional* (Mealy) and *unconditional* (Moore) outputs are all represented in

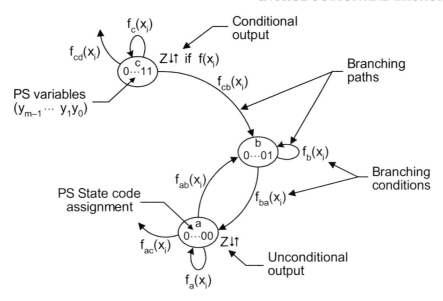

FIGURE 1.4: Fully documented state diagram as interpreted for an asynchronous FSM showing branching paths, branching conditions, conditional and unconditional outputs, and state code assignments.

the state diagram segment. The down/up arrows symbol ($\downarrow\uparrow$) is used to indicate an active output upon entering a state and an inactive output upon leaving the state, subject to any conditional input indicated. It also serves to visually distinguish outputs from inputs. A branching path out of and back into a given state symbol is called a *holding condition*. In Figure 1.4, the holding condition for state a is denoted as $f_a(x_i)$ or, alternatively, $f_{aa}(x_i)$, and similarly for states b and c. All others are either out-branching or in-branching paths as indicated by the sense of the arrows. Note that the branching conditions $f_{\alpha\beta}(x_i)$ given in Figure 1.4 are the input literals required to execute a given transition from state α to state β, or from state α to state α. Thus, $f_{\alpha\beta}(x_i)$ can represent any number of literals (inputs) as, for example, X, $\overline{ST}$, Add $\cdot$ $\overline{V}$, or $B + \overline{\text{Hold} \cdot C}$, etc.

The reader will learn with practice that the fully documented state diagram is by far the easiest, most lucid, and visually satisfying means of representing the sequential behavior of an FSM. In this regard, it is important to note that we have deliberately opted not to use *burst mode* and *extended burst mode* FSM label notation in state graph construction (see Glossary for definitions) because, to do so, would defeat the underlying purposes of this text. It will become clear to the reader that the fully documented state diagram is a very powerful tool when used to identify and eliminate a host of timing defects in asynchronous FSMs. However, to use the fully documented state diagram for such purposes, two rules must "ordinarily" be followed in its proper construction. The first of these rules is called the *sum rule* stated as follows:

Sum Rule

The Boolean sum of all branching conditions from a given state must be logic 1.

Applied to state a in Figure 1.4, the sum rule would be satisfied iff $f_a + f_{ab} + f_{ac} = 1$. Suppose that the branching conditions from state a are $f_a = XY$, $f_{ab} = \bar{X}$ and $f_{ac} = X\bar{Y}$. Clearly, the sum rule holds under these branching conditions, because their Boolean sum would be logic 1 (see Appendix A.2 for a review of Boolean algebra). An example of failure to meet the sum rule would be if $f_{ab} = \bar{X}Y$, the others remaining the same. Mathematically, the sum rule can be expressed for all j-to-i transitions as

$$\sum_{i=0}^{n-1} f_{i \leftarrow j} = 1 \tag{1.3}$$

where $f_{i \leftarrow j}$ represents the branching condition from the jth state to each ith summed over n states.

If the sum rule is not satisfied, one or more of the branching conditions would not be accounted for and failure of the FSM could result. However, there are occasions in the design of an asynchronous FSM where it is desirable not to have the sum rule apply. In these special cases, certain input conditions are never allowed thereby permitting the FSM to function normally in violation of the sum rule. An example in this case would be if $f_a = XY$, $f_{ab} = \bar{X}Y$, and $f_{ac} = X\bar{Y}$. In this case, the input condition $\bar{X}\bar{Y}$ is not accounted for. Thus, the FSM would only operate properly if the input change $XY \rightarrow \bar{X}\bar{Y}$ is never permitted to occur.

A second rule is called the *mutually exclusive requirement*. Although branching accountability is met via the sum rule, the asynchronous FSM may still malfunction if two or more of the branching conditions from a given state "overlap." This rule may be stated as follows:

Mutually Exclusive Requirement

Each possible branching condition from a give state must control no more than one branching path.

Mathematically, this may be expressed as

$$f_{i \leftarrow j} = \overline{\sum_{\substack{k=0 \\ k \neq i}}^{n-1} f_{k \leftarrow j}} \quad \text{or} \quad \left(f_{i \leftarrow j} \right)\left(f_{k \leftarrow j} \right) = 0 \atop \text{for all } i \text{ and } k, \text{ iff } k \neq i \tag{1.4}$$

where each branching condition is found to be the complement of the Boolean sum of those remaining. Applied to state a in Figure 1.4, Eqs. (1.4) requires that

$$\left\{\begin{array}{lll} f_a = \overline{f_{ab} + f_{ac}} & \text{or} & f_{ab} \cdot f_{ac} = 0 \\ f_{ab} = \overline{f_a + f_{ac}} & \text{or} & f_a \cdot f_{ac} = 0, \quad \text{etc.} \end{array}\right\} \qquad (1.5)$$

This example demonstrates that a simpler way of satisfying Eqs. (1.4) would be to AND each pair of branching conditions from a given state. If the result is logic zero for *all* ANDed pairs, then all branching conditions are mutually exclusive. Consider that $f_a = \overline{X}\overline{Y}$, $f_{ab} = X$, and $f_{ac} = Y$ in Figure 1.4. It is easily seen that the sum rule is obeyed but the mutually exclusivity condition is violated, because X and Y both contain XY. Thus, if the holding condition in state a is $\overline{X}\overline{Y}$ and the change $\overline{X}\overline{Y} \rightarrow XY$ occurs, then branching to either state b or state c can occur leading to a possible unresolved condition that can result in an error transition (or metastability) and failure. Notice also that $f_{ab} \cdot f_{ac} = X \cdot Y = XY$ does not satisfy the null product result required by Eqs. (1.4). Of course, if branching condition XY is never permitted to occur, then there is no potential problem.

1.6   THE MAPPING ALGORITHM

Fortunately, there is a simple means of designing and analyzing asynchronous FSMs having any specified memory stage, and a simple means of converting back and forth between memory elements. This simple means begins with what is called the *mapping algorithm*. There are three steps that must be followed before applying the mapping algorithm. They are:

1. Select the FSM to be designed and represent it in the form of a fully documented state diagram. The output (OP) logic can be mapped and obtained at this time.
2. Select the memory element (LPD or an SR type) and represent this memory element in the form of an excitation table.
3. Plot the NS EV Karnaugh maps (K-maps) from items (1) and (2) by using the mapping algorithm below, and loop out a minimum or near-minimum NS logic from the EV K-maps. The use of computer-aided minimization software may be necessary to complete this step.

Mapping Algorithm for FSM Design

AND each memory input logic value in the excitation table with the corresponding EV branching condition (or canonical value) in the state diagram for the FSM to be designed, and enter the result in the appropriate cell of the NS K-map.

1.7 APPLICATION OF THE MAPPING ALGORITHM TO SIMPLE LPD MODEL DESIGNS

Remember that we must begin by following the three preliminary steps cited above before applying the mapping algorithm. First, it is necessary to understand what set and reset mean logically. Simply put, Set refers to a $0 \rightarrow 1$ transition and Reset to a $1 \rightarrow 0$ transition. These can be represented by a simple two-state state diagram as shown in Figure 1.5a, where the sum rule must hold for state 0 and for state 1 taken separately. Thus, $\bar{S} + S = 1$ and $R + \bar{R} = 1$. Now, we AND the logic 1 (for Set) in Figure 1.5b with the branching condition S for the $0 \rightarrow 1$ transition in Figure 1.5a and place the result (S) in the PS state 0 cell of the EV K-map given in Figure 1.5c. Similarly, we AND logic 1 (for Set Hold) with the holding condition $\bar{R}$ in state 1 and enter the result ($\bar{R}$) in the PS state 1 of the EV K-map. ANDing the branching conditions $\bar{S}$ and R with logic 0, as required by the excitation table in Figure 1.5b, results in null entries in the NS K-map of Figure 1.5c.

The result in Figure 1.5c is important because it permits K-map conversion, back and forth between the LPD and SR models, leading to a design conversion between the two models. In effect, this amounts to *memory conversion*. For reference purposes, we state this result again here

$$Y = \bar{y}S + y\bar{R} \qquad (1.6)$$

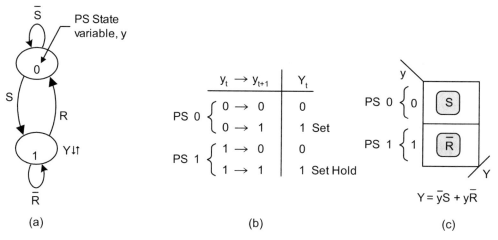

FIGURE 1.5: LPD-to-SR memory conversion. (a) The SR state diagram. (b) Excitation table for memory in the LPD model as given by Figure 1.2b. (c) EVK-map and LPD-to-SR memory conversion logic derived from the state diagram in (a) together with the excitation table in (b) and the use of the mapping algorithm.

The utility of conversion between the LPD and SR models is that it is generally easier to obtain the NS logic for the LPD model than for the SR model. Thus, conversion to the SR model via K-map conversion can save time and reduce errors, and the result applies to either the use of nested basic cells or Muller C-elements. The use of a minimization software that accepts map EVs, such as BOOZER (see Preface), further reduces the time and effort required in obtaining reliable results. All of this will be covered at the appropriate time.

To further help the reader understand the mapping algorithm as applied to the LPD model design process, consider the simple three-state FSM in Figure 1.6a. Here, a three-state, fully documented state diagram is shown having two state variables (y_1 and y_0), two external inputs (A and B), and a single conditional output Z. The output must go active in state 00 but only if A is active. The FSM will hold in state a under input conditions $A + \bar{B}$ or any logic combination of inputs contained in $A + \bar{B}$, that is, $\bar{A}\bar{B}$, $A\bar{B}$ and AB. Notice that the sum rule and mutual exclusivity requirement hold for all three states.

Shown in Figure 1.6b is the *state table* (flow table), which is the tabular form precisely representing the state diagram but more amenable to computer-aided design. The arrows represent the possible transitions. The input axis AB is unfolded in 2-bit Gray code with input domains A and B indicated by brackets. Each cell entry is a state identifier representing the specific state code

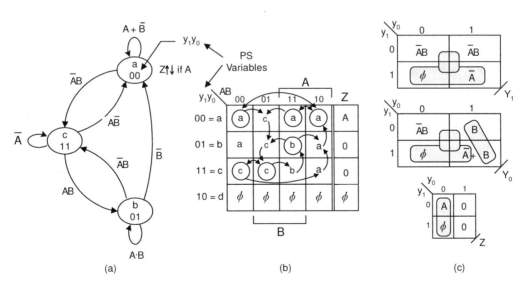

(a)   (b)   (c)

FIGURE 1.6: LPD design of a simple three-state asynchronouc FSM. (a) Fully documented state diagram. (b) State table. (c) NS and output K-maps showing optimal cover (shaded) loops.

assignment shown on the vertical axis of the state table in agreement with the state diagram. State identifiers that are encircled represent holding conditions that satisfy both the state diagram and the stability criteria of Eqs. (1.1) and (1.2). In an asynchronous FSM, this means that the FSM is stable in any state for which the stability criteria is satisfied, but is unstable otherwise and must transit.

As an example, the FSM in Figure 1.6a is stable in state b if both inputs A and B are simultaneously active. Then, for the FSM to transition $b \rightarrow a$, B must go inactive; hence, $A\bar{B}$. The transitions $00 \rightarrow 11$ ($a \rightarrow c$) and its reverse $11 \rightarrow 00$ ($c \rightarrow a$) require that the FSM must transit via either state 01 or the don't care state 10 designated ϕ_{10}. Thus, the $00 \rightarrow 11$ transition under input conditions $\bar{A}B$ can be guaranteed only if transitions from $01 \rightarrow 11$ and $10 \rightarrow 11$ contain $\bar{A}B$, which they do. Similarly, the $11 \rightarrow 00$ transition under input conditions $A\bar{B}$ can be guaranteed only if the transitions from $01 \rightarrow 00$ and $10 \rightarrow 00$ contain $A\bar{B}$, which they do. The fact that the transitions from missing state 10 contain the appropriate branching conditions will become clear when we discuss asynchronous FSM analysis in Chapter 7. The transitions $00 \rightarrow 11$ and $11 \rightarrow 00$ via state 01 or 10 are examples of *permissible race conditions* to be discussed in Chapter 3.

The K-maps in Figure 1.6c are generated by combining application of the LPD excitation table (the characterization of the memory) in Figure 1.2b with the state diagram in Figure 1.6a by using the mapping algorithm. This is accomplished by taking each state variable in turn (e.g., y_1 first then y_0) and plotting the NS EV K-maps for $0 \rightarrow 1$ and $1 \rightarrow 1$ SET transitions between states. Beginning with state 00, there is only one set condition for the y_1 state variable and that is a $0 \rightarrow 1$ transition to state c under branching condition $\bar{A}B$. Thus, $\bar{A}B$ is placed in cell 00 of the Y_1 NS variable K-map. Continuing to state 01, there is again only one set condition for the y_1 state variable, a $0 \rightarrow 1$ transition to state c under branching condition $\bar{A}B$, which is placed in the cell of coordinates 01 of the Y_1 NS variable K-map. In state 11, the $1 \rightarrow 1$ set-hold occurs under input condition $\bar{A}$, which is placed in the 11 cell of the Y_1 K-map. State 10 is a don't care state, so a ϕ is placed in the 10 cell. Now consider the y_0 state variable and state 11. The y_0 state variable transits from state 11 to 11 under input condition $\bar{A}$ but also from state 11 to 01 under input conditions AB. Therefore, the Boolean sum of these two set conditions for state variable y_0 is, by the absorptive law in Appendix A.2, $\bar{A} + AB = \bar{A} + B$, which must be placed in the 11 cell of the Y_0 K-map. (Note: Always minimize any combined branching conditions.) Then, for state 01 the transitions $01 \rightarrow 01$ and $01 \rightarrow 11$ require that $AB = \bar{A}B = B$ be placed in the 01 cell of the Y_0 K-map. Again, a don't care symbol ϕ is placed in the 10 cell of the Y_0 K-map. What is the 00 cell entry for the Y_0 K-map?

The output K-map for Z requires that an A be placed in the 00 cell because state 00 has a conditional output (Z if A), meaning that this output can go active in state 00 only if A is active independent of input B. The other two states, 01 and 11, have no output assigned to them so logic 0 must be placed in cells 01 and 11 of the Z K-map. Cell 10 must also contain a don't care symbol, ϕ.

When the mapping process has been completed by following Appendix A.3, the maps are read (looped out) to give

$$Y_1 = \bar{A}B + y_1\bar{A} \qquad Y_0 = \bar{A}B + y_0B + y_1\bar{A} \qquad Z = \bar{y}_0A \qquad (1.7)$$

where *prime implicants* (PIs) $\bar{A}B$ and $y_1\bar{A}$ are shared between NS variables Y_1 and Y_0, and are called *shared PIs*. This yields a total *gate/input tally* of 6/13 excluding inverters that are used in the LPD logic circuit. To help understand the concept of gate/input tally and application of the LPD model, the NS and output logic in Eqs. (1.7) are shown in Figure 1.7 to be implemented with a NAND/INV logic circuit by using *mixed-logic notation*. Here, as in the LPD model of Figure 1.1, fictitious memory elements are represented by the symbol M_i separating the NS variables Y_i from the PS variables y_i. The circuit is implemented assuming that the inputs arrive active high and that the output Z is issued active low (a matter of choice). The forward slash on a gate input line indicates a *logic incompatibility* requiring that the input to that logic gate be complemented in the gate output. It is not obvious, but the shared PI $y_1\bar{A}$ provides a transition path from state 00 to state 11 via don't care state 10 should the FSM transit by that path. This will be made clear when FSM analysis is discussed in Chapter 7. Note that Figure 1.7 verifies that the gate outputs agree with the p-terms in Eqs. (1.7) and that the total gate/input tally is 6/13 exclusive of inverters. Once the reader is familiar with LPD model designs, the fictitious memory elements can be removed but implied. Thus, in this case, the state variable outputs will then be given as $y_1(H)$ and $y_0(H)$.

1.7.1 Mixed-Logic Notation and the Cardinal Rule

Learning to construct logic circuits in mixed-logic notation provides the designer with some significant advantages, so a brief review of mixed-logic notation will be useful at this time. First, notice

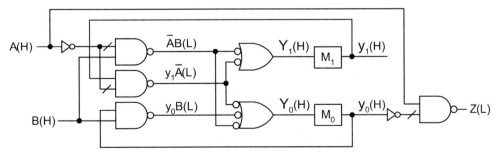

FIGURE 1.7: Implementation of Eqs. (1.7) with NAND/INV logic assuming that the inputs arrive active high and that the output is issued active low. (The slash indicates a logic incompatibility.)

that the inputs and outputs of the logic circuit in Figure 1.7 carry what are called *activation level indicators*. They are:

(*H*) signifies ACTIVE HIGH where Logic 1 is the ACTIVE state
(*L*) signifies ACTIVE LOW where Logic 0 is the INACTIVE state

The DeMorgan relations for any logic function α, and relationships between (H) and (L) and the high and low voltage levels (HV and LV), are essential to understanding the mixed-logic notation and design methods in the *logic domain* used in this text. These are, respectively:

$$\left\{\begin{matrix} \alpha(L) = \bar{\alpha}(H) \\ \alpha(H) = \bar{\alpha}(L) \end{matrix}\right\} \quad \left\{\begin{matrix} 1(H) = 0(L) \text{ corresponds to HV} \\ 0(H) = 1(L) \text{ corresponds to LV} \end{matrix}\right\} \tag{1.8}$$

The simplicity and utility of this type of notation will become apparent with practice. The above notation will be used extensively throughout this text. Appendix A, provided at the end of this text, will review mixed-logic symbology and other fundamentals that the designer will find useful. For now, the cardinal rule to be followed can be stated as follows:

Cardinal Rule

Always design or analyze a logic circuit in mixed-logic notation and symbology. Use of positive logic or voltage-level notation must be left to the hardware implementation stage.

Following this cardinal rule can help users avoid numerous errors and failure. The reader will learn that asynchronous state machine design and analysis is complex, requiring a simplified notation that maximizes the probability of success. Use of mixed-logic notation and symbology is superbly suited to accomplishes that. Readers who have been taught only positive logic must unlearn the positive logic notation and aggressively adapt to the cardinal rule.

1.8   DESIGN OF BASIC MEMORY ELEMENTS AND THEIR CHARACTERISTICS

Before we can delve into the subject of the nested element models represented in Figure 1.3, we must first discuss the nature of the SR nested elements that constitute the memory for these models. As mentioned earlier, there are two classes of memory elements: (1) the basic cells classified as set-dominant and reset-dominant that operate in the fundamental mode, and (2) the Muller C-elements that operate outside the fundamental mode. We will discuss their design and characteristics in turn.

1.8.1 Basic SR Cells

Shown in Figure 1.8 is the design of the *set-dominant* basic memory cell, which is itself the simplest type of gate-level asynchronous FSM possible. Figure 1.8a shows the *operation table* for the set-dominant basic cell and Figure 1.8b depicts the state diagram derived from the operation table. Notice that in the operation table the SR conditions 10 and 11 both require a set condition, hence the name set-dominant. The 00 condition requires a reset hold or set hold in states 0 or 1, respectively. The only reset condition occurs for $\bar{S}R$, meaning that S is inactive when R is active. Also note that the sum rule and mutual exclusivity condition are both satisfied for the state diagram.

Combining the state diagram in Figure 1.8b with the excitation table in Figure 1.8c for the LPD model by using the mapping algorithm results in the NS EV K-map shown in Figure 1.8d. The optimum cover, looped out of the EV K-map in *minterm code*, gives the sum-of-products

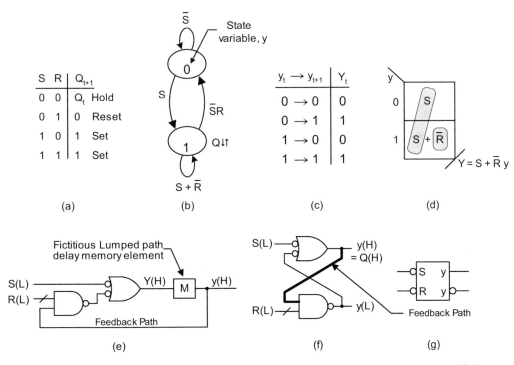

FIGURE 1.8: Design of the set-dominant basic memory cell by using the LPD model. (a) Operation table. (b) State diagram. (c) Excitation table for the LPD model. (d) NS K-map and minimum cover. (e) NAND Logic circuit showing the fictitious LPD memory element and the single feedback path. (f) Logic circuit with fictitious memory element removed. (g) Logic symbol.

logic expression $Y = S + \bar{R}y$, which is implemented in mixed-logic form with and without the fictitious memory element in Figure 1.8e and 1.8f, respectively. Notice that this simple NAND centered
asynchronous FSM has but one feedback path. The appearance of the "cross-coupled" NAND gates
in Figure 1.8f does not change this fact. The logic circuit symbol, given in Figure 1.8g, provides
a suitable representation of the set-dominant basic cell that can be used in circuits for which the
nested element model in Figure 1.3 is applied. Thus, the active low indicator bubbles on the inputs
circuit symbol in Figure 1.8g imply active low S and R inputs.

Shown in Figure 1.9 is the LPD design of the reset-dominant basic cell. The state diagram
for this cell in Figure 1.9b is derived from its operation table in Figure 1.9a. It is mapped in Figure 1.9d by using the mapping algorithm to combine the state diagram with the excitation table
for the LPD model given in Figure 1.9c. The K-map is looped out in *maxterm code* to yield the

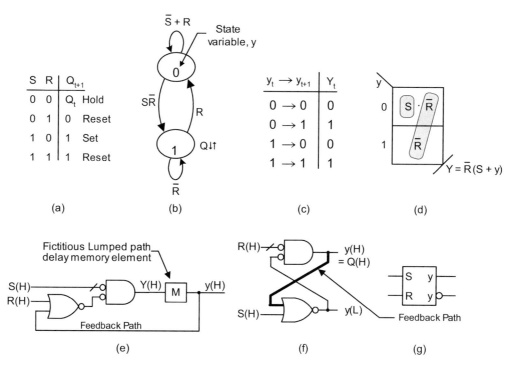

FIGURE 1.9: Design of the reset-dominant basic memory cell by using the LPD model. (a) Operation
table. (b) State diagram. (c) Excitation table for the LPD model. (d) NS K-map and minimum cover.
(e) NOR Logic circuit showing the fictitious LPD memory element and a single feedback path.
(f) Logic circuit with the fictitious memory element removed. (g) Logic symbol.

product-of-sums (POS) logic expression $Y = \bar{R}(S + y)$. In maxterm code K-map minimization, the domains are complemented and read as POS. Appendix A.3 reviews EV K-map minimization but in minterm code only. (See Tinder's book in Endnotes for an exhaustive treatment of EV K-map mapping minimization.)

The logic circuit for $Y = \bar{R}(S + y)$ is implemented in NOR logic with and without the fictitious memory element as shown in Figure 1.9e and 1.9f. Again, we observe that there is but one feedback path. Notice that for the reset-dominant basic cell the inputs arrive active high (H), whereas for the set-dominant (NAND-centered) basic cell they arrive active low (L). A suitable logic symbol for the reset-dominant cell without active low indicator bubbles is given in Figure 1.9f. An inspection of the state diagram in Figure 1.9b indicates that the sum rule and mutual exclusivity condition are both satisfied.

The logic character of the basic memory cells is best understood by the use of *timing diagrams* (logic waveforms). Shown in Figure 1.10 are the combined timing diagrams for the set- and reset-dominant basic cells that have been taken from a logic simulator described in the Preface but dressed up with a drawing tool. To save space, the input waveforms are made to appear exactly the same for the two basic cells but with the understanding that their activation levels SL,SH and RL,RH must be in agreement with their respective figures, Figures 1.8f and 1.9f. Thus, the inputs to the cross-coupled

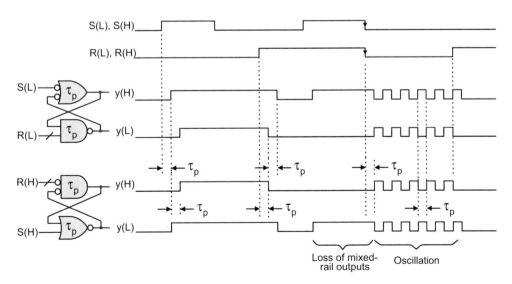

FIGURE 1.10: Timing diagrams for the set-dominant and reset-dominant basic cells showing loss of mixed-rail outputs for the $S,R = 1,1$ condition, and the oscillatory behavior that results when S and R change $1 \rightarrow 0$ simultaneously.

NAND cell (set-dominant) are SL and RL, whereas for the cross-coupled NOR cell (reset-dominant) they are SH and RH.

Outputs $y(H)$, $y(L)$ are called *mixed-rail outputs* because they normally issue either as 1H,1L or as 0H,0L—in the physical domain, the voltages are the inverse of one another, according to Section 1.7.1. However, these outputs do not change simultaneously but rather their changes are separated by the *propagation delay*, τ_p, of a single gate, which has arbitrarily been taken to be the same for both NAND and NOR gates. Notice in Figure 1.10 that the $y(L)$ output from the cross-coupled NAND gates is symmetrically set inside of the $y(H)$ output by time delays of τ_p. Conversely, for the cross-coupled NOR gates, the $y(H)$ output is symmetrically set inside of the $y(L)$ output, again by time delays denoted by τ_p.

Under certain conditions, the two basic cells shown in Figure 1.10 can *lose their mixed-rail output character*. Loss of mixed-rail output character means that both outputs go active high, 1H, as in the NAND-based cell, or they both go active low, 0H, as in the NOR-based cell. Thus, physically these outputs are issued at the same voltage level, HV for the NAND-based cell and LV for the NOR-based cell. Remember that $1(H) = 0(L)$ and $0(H) = 1(L)$ in mixed-logic notation. When both inputs to either basic cell go active and then transition $1 \rightarrow 0$ simultaneously as shown in Figure 1.10, the basic cell may become metastable and either "hang up" in a state that is neither a set not reset, or it may oscillate. This supports the need to avoid the $S,R = 1,1$ condition when using basic cells as memory elements and the need to operate them in the fundamental mode, where input changes must be minimally separated in time.

If a set of basic memory cells are to be used as memory elements in an asynchronous logic circuit, in agreement with the nested element model of Figure 1.3, they must be characterized by an excitation table. To design an asynchronous FSM, the mapping algorithm requires that the state diagram for the FSM (to be designed) be combined with the excitation table for the selected memory element via the mapping algorithm. Shown in Figure 1.11 are the excitation tables for the set- and reset-dominant basic cells as derived from their respective state diagrams. We notice from the state diagrams that the $S,R = 1,1$ condition exists under Set ($0 \rightarrow 1$) and Set Hold ($1 \rightarrow 1$) changes for the set-dominant basic cell, and under Reset Hold ($0 \rightarrow 0$) and Reset ($1 \rightarrow 0$) for the reset-dominant basic cell. Thus, $S,R = 1,1$ is inherent in these two basic cells.

The excitation tables in Figure 1.11 can be combined (merged) to eliminate the presence of the $S,R = 1,1$ condition. By taking from Figure 1.11b and 1.11c only those SR entries that are enclosed in shaded loops, a combined form of the excitation table results in the absence of the $S,R = 1,1$ condition. Thus, a *generic* form of the excitation table results that is applicable to either the set- or reset-dominant basic cell. Use will be made of the combined form in Section 1.9 and throughout the text. Note that each input in an ORed holding condition is taken separately and shown in parentheses. Thus, $S + \bar{R}$ for state *1* in Figure 1.11a is presented as ($1\ \phi$) for S and ($\phi\ 0$) for $\bar{R}$.

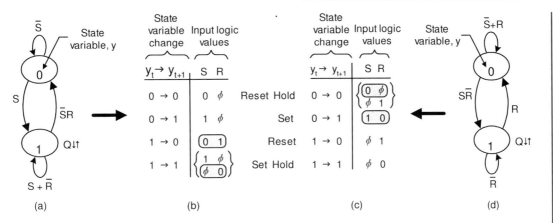

FIGURE 1.11: Excitation tables for the basic memory cells. (a) State diagram for the set-dominant basic cell derived from the operation table in Figure 1.8a. (b) Excitation table for the set-dominated basic cell derived from its state diagram in (a). (c) Excitation table for the reset-dominated basic cell derived from its state diagram in (d) or from its operation table in Figure 1.9a. (Note that the SR entries within shaded loops are those shared in common between the two excitation tables.)

1.8.2 Muller C-Elements

Shown in Figure 1.12a is the transistor circuit for a normal Muller C-element—there is no traditional logic gate representation as is the case of the basic cells. Whereas basic cells have a single feedback path to the NS forming logic, the Muller C-element has only cross-coupled inverters in its output with a *weak (keeper) inverter* that serves as the memory. Consequently, a C-element operates outside of the fundamental mode and cannot transition $0 \to 1$ until both inputs go active, nor can it transition $1 \to 0$ until both inputs go inactive. The truth table in Figure 1.12b derives from the transistor circuit, and the state diagram in Figure 1.12c is constructed directly from the truth table. Both represent the sequential behavior of the C-element. The two inputs can rendezvous active anytime to execute the $0 \to 1$ transition or can arrive inactive anytime to produce the $1 \to 0$ transition; otherwise, the C-element must hold in its respective state. No adverse effects will ensue from simultaneous input changes, in contrast to the simultaneous $1 \to 0$ transitions in basic cells. The symbol in Figure 1.12d serves as a suitable logic circuit symbol for a normal C-element and will be used as such throughout this text. The clear (CL) feature allows the C-element output $y(H)$ to reset at $0(H)$ when $CL = 1(L)$ and to allow normal operation following a $CL = 0(L)$.

Figure 1.13 shows the four representations for a *complimentary C-element* with clear. Now, the inputs must rendezvous with complementary activation levels (one active high, the other active low) before a transition $0 \to 1$ or $1 \to 0$ is executed. The truth table and state diagram in Figure

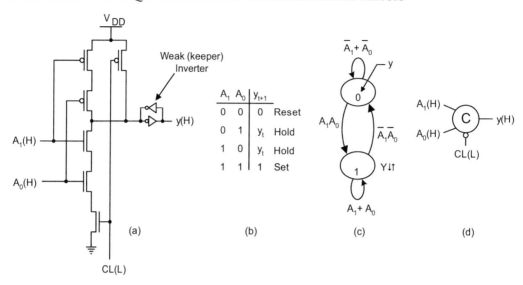

FIGURE 1.12: The normal Muller C-element. (a) Transistor circuit with clear (CL). (b) Positive logic truth table representation of the transistor circuit. (c) State diagram derived from the truth table. (d) Logic circuit symbol.

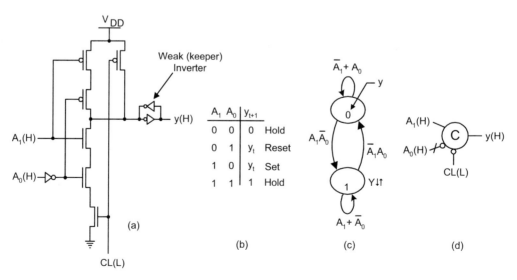

FIGURE 1.13: The complementary Muller C-element with clear (CL). (a) Transistor circuit showing inverter on the A_0 input. (b) Positive logic truth table representation of the transistor circuit. (c) State diagram derived from the truth table. (d) Logic circuit symbol.

1.13b and 1.13c convey this requirement. The circuit symbol in Figure 13d caries an *active low indicator bubble* representing *logic level conversion*. The clear feature is again added by following the configuration shown in Figure 1.12a. With regard to inverters, the reader must remember the following:

Function of an Inverter

An inverter does not invert a voltage signal in the logic domain but merely converts positive logic to negative logic or vice versa—it is the physical inverter that inverts voltage levels.

Note that the C-element in either Figure 1.12 or 1.13 can be designed without a clear (CL) input by shorting the NMOS to ground and by removing the PMOS (an open circuit at the PMOS).

1.9   SUMMARY OF THE EXCITATION TABLES

An excitation table is the means by which we characterize the memory to be used in the design of a logic circuit. The excitation table is combined with the state diagram, for the FSM to be designed via the mapping algorithm to produce the NS EV K-maps. Examples of this important process have been previously given in Figures 1.5 and 1.6. Shown in Figure 1.14 are the important excitation tables used in this text for the LPD model and the nested element models given in Figures 1.1 and 1.3, respectively. Column 1.14(b) is the same as that in Figure 1.2 used for the LPD model. Column

	LPD	Complementary C-element		Combined Basic Cells		Normal C-element	
$y_t \to y_{t+1}$	Y	A_1	A_0	S	R	A_1	A_0
Reset Hold $0 \to 0$	0	$\begin{cases} 0 \\ \phi \end{cases}$	$\begin{cases} \phi \\ 1 \end{cases}$	0	ϕ	$\begin{cases} 0 \\ \phi \end{cases}$	$\begin{cases} \phi \\ 0 \end{cases}$
Set $0 \to 1$	1	1	0	1	0	1	1
Reset $1 \to 0$	0	0	1	0	1	0	0
Set Hold $1 \to 1$	1	$\begin{cases} 1 \\ \phi \end{cases}$	$\begin{cases} \phi \\ 0 \end{cases}$	ϕ	0	$\begin{cases} 1 \\ \phi \end{cases}$	$\begin{cases} \phi \\ 1 \end{cases}$
(a)	(b)	(c)		(d)		(e)	

FIGURE 1.14: Summary of the excitation tables. (a) PS-to-NS transitions. (b) LPD model. (c) Complementary C-elements. (d) Combined form for SR basic cells (e) Normal C-elements.

1.14(c) is valid for the complementary C-element in Figure 1.13, column 1.14(d) applies to the combined basic cells (either a set- or reset-dominant) as in Figure 1.11, and column 1.14(e) applies to a normal C-element. The reader will notice the similarity between entries for the complementary C-elements and those for the combined basic cells illustrated by loops. Equation (1.6) gives not only the conversion logic between the LPD model and basic cells but also the conversion logic between the LPD model and complementary C-elements. The idea here is to first design with the LPD model, which is relatively easy, and then convert to either a C-element or basic cell memory design.

1.10   HUFFMAN VS. MULLER ASYNCHRONOUS FSMs

Historically, the Huffman and Muller schools of thought derive from and are credited to the early work of D. A. Huffman and D. E. Muller. The general requirement of a Huffman circuit is that the input changes must be minimally separated in time—that is, the circuit must operate in the fundamental mode. In contrast, the Muller circuit requires a "ready" signal generated by the circuit indicating that another input change is permitted. Thus, the Muller circuits are often referred to as *speed-independent circuits* because they operate outside of the fundamental mode. Over time, these definitions have been blurred somewhat by various interpretations. However, they are adequate for our purposes. We will concentrate on the proper design and analysis of FSMs whose character may be classified as either Huffman or Muller type, but this will be done usually with some qualification.

The use of C-element memory elements in an asynchronous FSM could seemingly be thought of as a quasi-Muller circuit design because a C-element itself does not have feedback to its NS logic as is required for the fictitious LPD and nested basic SR cell memory elements. So although a C-element memory operates outside of the fundamental mode, its use as memory in an asynchronous FSM does not completely satisfy the requirements for a Muller circuit design. Of course, such a design lacks a "ready" signal to enable the next input change, but the rendezvous character of the C-element makes this special signal partially a nonissue. The problem here, of course, is that outputs from C-elements are PS variables that are fed back as inputs to the NS forming logic for the FSM. In Chapters 8 and 9, we will deal with asynchronous systems that do operate so as to issue "ready" signals and rightfully deserve to be called Muller-type systems. In any case, the classification distinction of a Huffman vs. Muller design is an option left largely to the designer.

CHAPTER 2

Simple FSM Design and Initialization

In this chapter we will design a simple asynchronous finite state machine (FSM) by using both the lumped path delay (LPD) and nested C-element models. Conversion between these models will require an extension of the mapping algorithm given in Section 1.6. The reader will find that conversion between these models is, in effect, a quasi-conversion between Huffman and Muller designs—a powerful tool for use by a knowledgeable designer.

2.1 THE EXTENDED $Y \rightarrow SR$ MAPPING ALGORITHM

The extended mapping algorithm is inherent in Eq. (1.6) taken together with the excitation tables in Figure 1.14. For the convenience of the reader, we restate Eq. (1.6) here:

$$Y = \bar{y}S + y\bar{R} \qquad (1.6)$$

For the next state (NS) K-map conversion, $Y \rightarrow SR$, Eq. (1.6) and the excitation tables in Figure 1.14 require the following four steps:

Four Steps to $Y \rightarrow SR$ Conversion

1. For all that *is not* the NS y-domain, transfer it directly to the same domain cells in the S K-map.
2. For all that *is* the NS y-domain, transfer it complemented to the same domain cells in the R K-map.
3. Fill in each empty cell with a don't care (ϕ) ANDed with the complement of the S or R K-map entry of the "other" K-map having the same cell domain number.
4. Loop out the resulting K-maps to yield the optimum NS SR logic cover.

2.1.1 Application to FSM Design with C-Elements

Shown in Figure 2.1 are the $Y \rightarrow SR$ conversion NS K-maps and the optimum covers that result from following the four steps given in Section 2.1, as applied to the FSM in Figure 1.6. Note that a don't care ϕ ANDed with an entry means that the entire entry is nonessential: *Use it if you can, but*

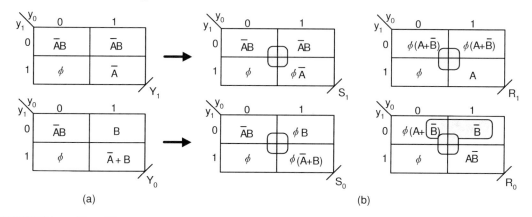

FIGURE 2.1: $Y \rightarrow$ SR conversion for the FSM in Figure 1.6a. (a) Original NS K-maps reproduced from Figure 1.6b. (b) Converted S,R EV K-maps showing optimum NS logic cover.

don't use it unless it is necessary to extract optimum or near optimum cover. Brief reviews of EV K-map minimization and incompletely specified functions are given in Appendix A.3.

For comparison purposes, the optimum NS and output forming logic for the LPD and SR models are given in Eqs. (1.7) and (2.1), where the NS SR logic results from covers given by the shaded loops in Figure 2.1b. An inspection of these equations indicates that the NS logic gate/input tally for the LPD model is 5/11 exclusive of inverters, whereas that for the nested element model yields a gate/input tally of 4/8 exclusive of inverters and C-elements.

$$\left.\begin{array}{c} Y_1 = \bar{A}B + y_1\bar{A} \\ Y_0 = \bar{A}B + y_0B + y_1\bar{A} \\ Z = \bar{y}_0 A \end{array}\right\} \quad \Rightarrow \quad \left.\begin{array}{cc} S_1 = \bar{A}B & R_1 = A \\ S_0 = \bar{A}B & R_0 = A\bar{B} + \bar{y}_1\bar{B} \\ Z = \bar{y}_0 A \end{array}\right\} \qquad (2.1)$$

The logic circuit for the LPD model result was previously given in Figure 1.7. The nested SR element model in Figure 1.3 and the SR logic in Eqs. (2.1) apply to either the use of basic cells or C-elements. However, an inspection of the excitation tables for the C-elements and basic cells in Figure 1.14 indicates there is an important difference in the way these equations are used. Complementation of the A_0 columns in Figure 1.14c and 1.14e satisfies Eq. (1.6) and allows $A_1 \rightarrow S$ and $A_0 \rightarrow R$ providing that the complementation requirement is followed:

Requirements for the *R* Inputs to C-elements and Basic Cells

To use C-elements in the nested element model, the *R* logic inputs MUST be complemented.
To use basic cells in the nested element model, the *R* inputs MUST NOT be complemented.

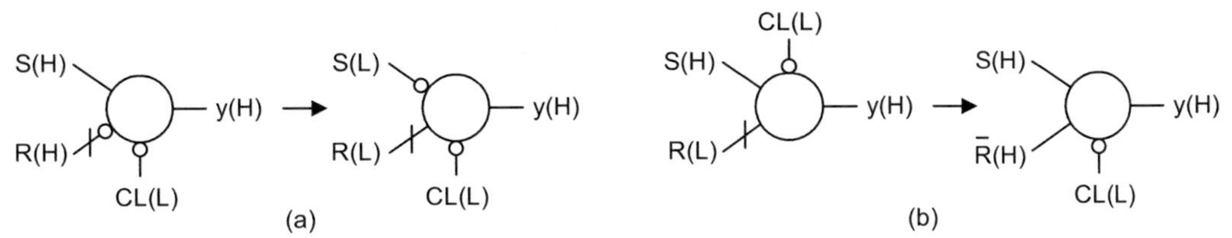

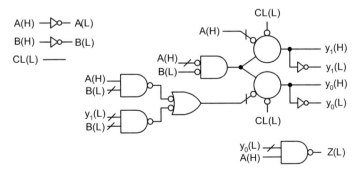

FIGURE 2.2: Typical examples of alternative uses of C-elements as required for SR logic. (a) Complimentary C-elements. (b) Normal C-elements.

This complementation requirement is best illustrated in Figure 2.2 for either the complementary or normal C-elements. Thus, having obtained the SR logic from the LPD model, the designer is free to choose which inputs are to serve as the S and R inputs owing to the transistor circuit configurations for the C-elements given in Figures 1.12 and 1.13.

To illustrate, the C-element-based logic circuit, representing the SR expressions in Eqs. (2.1), is shown in Figure 2.3. In this circuit, use is made of the *wireless connection feature* that is used to simplify the appearance of a logic circuit. It is the same feature that is recommended for use with the logic simulator described in the Preface. As a reminder, the *slash* appearing in the figure indicates a logic level incompatibility requiring that the input to a gate where it applies be complemented in the output of that gate. For example, the two-input conjugate NOR gate whose output is connected to the S inputs of the two C-elements has a mixed-logic output $\bar{A}B(H)$. Remember also that an inverter is a logic level converter, $(H) \rightarrow (L)$ or $(L) \rightarrow (H)$, and must not be given a physical meaning such as (high voltage) $\leftrightarrow$ (low voltage). These reminders are all part of the mixed-logic notation symbology used in this text to design or analyze logic circuits in the logic domain. Refer to Section 1.7.1 and Appendix A for details.

FIGURE 2.3: C-element logic design (with clear) of the same FSM featured in Figure 1.7 demonstrating the difference between the LPD and nested C-element model by using the wireless-connection feature.

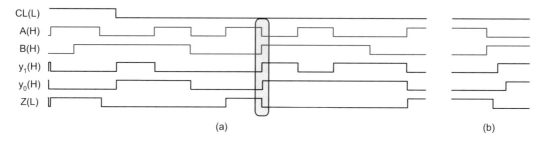

FIGURE 2.4: (a) Simulation of the C-element logic circuit in Figure 2.2 showing initialization with clear CL(*L*), its release, and the resulting transitions in agreement with the state diagram in Figure 1.6a. (b) Blowup view of the shaded region in (a) showing the transition from state 00 to state 11 via don't care state 10.

Simulation of the logic circuit in Figure 2.3 is given in Figure 2.4a, where use is again made of mixed-logic notation. The reader should follow the *A, B* input sequence to verify that the FSM simulation follows the sequential behavior required by the state diagram in Figure 1.6a. Notice that the circuit is initialized to the 00 state by setting the clear input CL(*L*) = 1(*L*) for a short period. Following release of the CL(*L*) signal, CL(*L*) → 0(*L*) =1(*H*), the reader should verify that the FSM transitions through the three states are consistent with Figure 1.6a. The transitions 00 → 11 and 11 → 00 follow paths via the state 01 or don't care state 10 as they must. This is demonstrated by the blowup in Figure 2.4b, where the transition $y_1(H)$, $y_0(H)$ = 00 → 11 follows a path through don't care state 10 spending a time in that state equal to the delay through the NOR gate plus a C-element shown in Figure 2.3. Simulation of the FSM in Figure 1.6a, the LPD model version of Figure 2.3, would look very similar to that in Figure 2.4. Naturally, requirements for the fundamental mode would have to be followed. The subjects of initialization and reset are discussed in Section 2.2.

2.2   INITIALIZATION OF ASYNCHRONOUS FSMs

Initialization of an FSM into a specific logic state is extremely important. A logic sequence must usually begin at an "origin" state and may terminate at the same state or at another state, or it may not terminate at all. The designer must remember that when power is applied to a circuit, random activity must not occur. Rather, the logic circuit must initialize stably into a specific state so that an orderly and predictable sequential behavior occurs. In an asynchronous FSM, this is especially important because the power-up action may involve switch "bounce," which could send the FSM into an undesirable and uncontrolled sequence of transitions. This is the reason why we introduce the concept of a *sanity circuit* that can avoid such problems and ensure that the FSM initializes into the required state and functions properly thereafter.

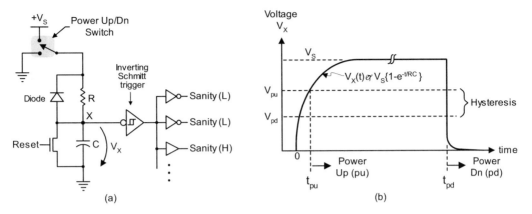

(a)   (b)

FIGURE 2.5: (a) Sanity circuit showing mixed-rail outputs. (b) *V-t* characteristic at node *X* for the sanity circuit in (a) showing power-up (V_{pu}) and power-down (V_{pd}) switching thresholds and hysteresis effect of the inverting Schmitt trigger.

2.2.1   Sanity Circuits—Design and Applications

FSMs can be initialized into any state, but are typically initialized into either a state of all 0's or all 1's. Furthermore, initialization can be done manually or by an analog circuit such as that shown in Figure 2.5a. This circuit is called a sanity circuit because its use is the only sane approach to initializing a logic circuit. In power down, the sanity signals are all active, $1(L)$ or $1(H)$, and the logic circuit to which the sanity signals are attached becomes initialized into the desired state. At power up and after a short period equal to the *time constant* for the sanity circuit, the sanity signals all go inactive, $0(L)$ or $0(H)$, and the logic circuit is enabled to function normally. The time constant is equal approximately to *RC* as indicated in Figure 2.5b. The CMOS implementation for an inverting Schmitt trigger is provided in Tinder's text (see Endnotes).

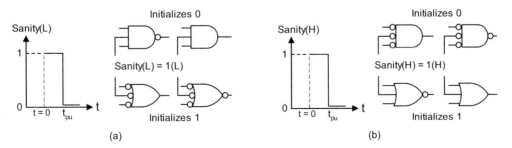

(a)   (b)

FIGURE 2.6: Gate requirements for initializing a logic 0 or a logic 1. (a) Active low gate input from the sanity circuit. (b) Active high gate input from the sanity circuit.

The gate requirements for initializing a logic 0 or logic 1 are given in Figure 2.6. Initialization by means of the NS logic is applicable to asynchronous FSMs that adhere to the LPD or nested basic cell models in Figures 1.1 and 1.3. Thus, the NS forming logic for such circuits can be implemented with either NAND, AND, NOR, or OR logic, and be initialized by a Sanity(L) signal according to Figure 2.6. Use of NAND logic to initialize a $0(H)$ and $1(H)$ is illustrated in Figure 2.7a and 2.7b, which is applicable to the LPD model with feedback paths to the NS logic. In contrast, the C-elements, shown in Figure 2.7c, initialize only to $0(H) = 1(L)$ when Sanity(L) $= 1(L)$ is applied. Naturally, it follows that an inverter on the C-element output would allow initialization to $0(L) = 1(H)$ if that is necessary. Obviously, initialization by means of say the NAND in Figure 2.7a and 2.7b requires additional NAND gate inputs that can slow down the circuit response to input change. Initialization by means of C-elements has an advantage because no additional inputs are required.

Now, the initialization of the FSM in Figure 2.3 can be easily understood. The CL(L) signal is the Sanity(L) that delivers a $1(L)$ for a short period, about equal to the time constant of the sanity circuit in Figure 2.5a. This drives the state variables y_1, y_0 to $0(H)$, which initializes the FSM into the 00 state as indicated in Figure 2.4. Then after a short period, the Sanity circuit goes to $0(L)$ and the FSM is enabled and function normally. In Figure 2.4, note also that when Sanity(L) = CL(L)

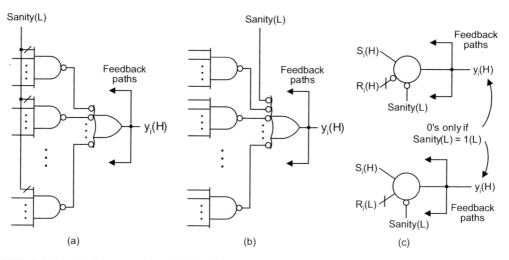

(a)   (b)   (c)

FIGURE 2.7: Initializing two-level NAND (SOP) logic and C-elements with Sanity(L). (a) NAND logic Sanity(L) $= 1(L) = 0(H)$, used to initialize a logic 0. (b) NAND logic with Sanity(L) $= 1(L)$ used to initialize logic 1. (c) Sanity(L) $= 1(L)$ inputs to C-elements as used to initialize only logic 0's.

goes to $0(L)$ and the inputs are $A(H) = 0$, and $B(H) = 1(H)$, the FSM transits $00 \rightarrow 11$ as it must according to the state diagram in Figure 1.6a. Because the output forming logic $Z(L) = \bar{y}_0 A(L)$ is not initialized, it is enabled to go active during initialization into the 00 state if A is active high. The output forming logic in Figure 2.3 can be initialized to $0(L)$ by using a three-input NAND gate with Sanity$(L) = 1(L)$ as one of the inputs.

CHAPTER 3

Detection and Elimination of Timing Defects in Asynchronous FSMs

There are five types of timing defects that, if present and active in asynchronous finite state machines (FSMs), can cause them to malfunction. Nearly all of these defects do not exist in synchronous FSMs because flip-flops, together with the sampling input called Clock, serve to filter them out. We will discuss these timing defects as they exist in asynchronous FSMs, indicating their potential to cause failure and examine the means by which they can be eliminated. Further discussion on asynchronous state machine design and analysis cannot occur until these timing defects are discussed in detail. The five types of timing defects are as follows:

1. Endless cycles (oscillations)
2. Critical races
3. Static hazards in the next state (NS) and output forming logic
4. Output race glitches (ORGs)
5. Essential hazards (E-hazards)

3.1 ENDLESS CYCLES

An *endless cycle* in an asynchronous FSM is nothing more than an *oscillation* operating at a frequency determined solely by the speed of the logic circuit involved in the production of the endless cycle. If this defect exists in the asynchronous FSM and is active, it turns the FSM into an oscillator and accordingly must be eliminated. Shown in Figure 3.1a is a general test used to detect an endless cycle involves ANDing the branching conditions between two states P and Q. If $f_{PQ} \cdot f_{QP} \neq 0$, then there exists an branching condition common between the two states that can cause the endless cycle. Such is the case in Figure 3.1b, where it is clear that $(A \oplus B) \cdot B = \bar{A}B$, revealing the common branching condition $\bar{A}B$ that can produce an uncontrolled oscillation. One possible means to eliminate this endless cycle involves simple changes in the branching conditions as shown in Figure 3.1c. Here,

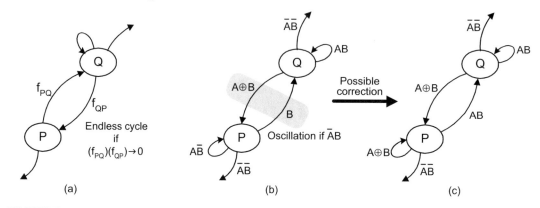

FIGURE 3.1: Endless cycles in asynchronous FSMs. (a) A seqment of a state diagram used as a model for endless cycle analysis. (b) Example of an endless cycle. (c) Elimination of the endless cycle in (b).

there now results $(A \oplus B) \cdot AB = 0$, thereby eliminating the endless cycles. Endless cycles can occur in a variety of state configurations including those involving more than two states.

3.2 RACES AND CRITICAL RACES

Any set of alternative paths from an origin state to a destination state, involving a change of two or more state variables, is called a *race* or *race path*. An example of race paths is illustrated in Figure 1.6a, where transitions between states 00 and 11 must transit via either state 01 or "don't care" state 10. Thus, race paths are not only permitted but in some cases are essential to the proper operation of the FSM. A race path that can result in a transition to, and stable residence in, an erroneous state is called a *critical race*. This timing defect must not be allowed to exist. The generalized state diagram segment in Figure 3.2a serves as a useful model for detection of race and critical race conditions associated with the transition between states P and Q under branching condition f_{PQ}. Simply stated, if f_{PQ} is contained in either holding condition f_R or f_S for race states R or S, as indicated in Figure 3.2c, then a critical race exists—the FSM can erroneously hold up indefinitely in such critical race states. The proper function of race states R or S must be to provide transition paths to destination state Q via branching condition f_{RQ} or f_{SQ}, as is the case in Figure 1.6a.

The formation and elimination of a critical race is illustrated in Figure 3.3. Here, it is seen that a transition $11 \rightarrow 00$ under branching condition $\bar{A}B$ involves the change in two state variables, $y_1 y_0$, and consequently must transit via either state 01 or 10. If the FSM should transit via state 10, it would be erroneously stuck in that state under holding condition B. Also, an output Z would be issued should input A go active while in state 10.

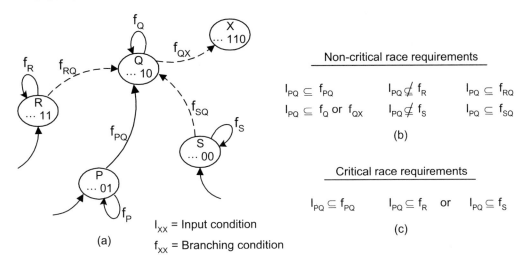

Non-critical race requirements

$$I_{PQ} \subseteq f_{PQ} \qquad I_{PQ} \nsubseteq f_R \qquad I_{PQ} \subseteq f_{RQ}$$
$$I_{PQ} \subseteq f_Q \text{ or } f_{QX} \qquad I_{PQ} \nsubseteq f_S \qquad I_{PQ} \subseteq f_{SQ}$$

(b)

Critical race requirements

$$I_{PQ} \subseteq f_{PQ} \qquad I_{PQ} \subseteq f_R \quad \text{or} \quad I_{PQ} \subseteq f_S$$

(c)

I_{XX} = Input condition

f_{XX} = Branching condition

(a)

FIGURE 3.2: Races and critical races in asynchronous FSMs. (a) Generalized state diagram segment used as a model for detection of races and critical races. (b) Requirements for non critical races. (c) Requirements for critical races.

The critical race shown in Figure 3.3 can be easily eliminated by using a correction path that diverts the 11 → 00 transition by way of state 01 under the same branching condition $\overline{A}B$. The transition is now 11 → 01 → 00, which is logically adjacent. This is called a *cycle* or *cycle path*, which, in this case, is permitted. Once the correction path is selected, the FSM must be redesigned by using either the lumped path delay (LPD) or nested element model. Further analysis must not continue until the FSM is free of endless cycles and critical races, all easily determined by inspection of the state diagram.

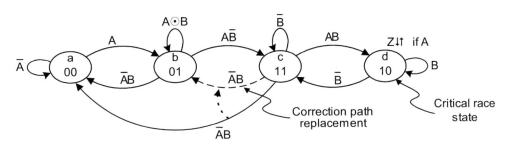

FIGURE 3.3: State diagram of an FSM showing a critical race via state 10 for the 11→00 transition and its elimination by providing a correction path represented by the dashed line.

3.3   STATIC HAZARDS IN THE NS AND OUTPUT FORMING LOGIC

A static hazard is a "glitch" (or instability) in an otherwise steady-state signal. A static hazard is produced by an input change that propagates along two asymmetric path delays through combinational logic consisting of inverters and gates. It is a *combinational hazard* called a *static hazard* (or S-hazard) even though it is quite dynamic in its transient behavior. Occurring in the NS logic of an asynchronous FSM, an S-hazard can cause malfunction of that FSM. For this reason, all static hazards in the NS forming logic must be eliminated. This subject will require the reader to be familiar with mapping and minimization techniques. To assist the reader, we categorize static hazards in the following manner:

1. A static-1 hazard occurs in sum-of-products (SOP) NS forming logic and is a $1 \rightarrow 0 \rightarrow 1$ transient (negative glitch) in an otherwise steady-state (static) logic 1 signal.
    (a) An *internally initiated static-1 hazard* is formed by a $1 \rightarrow 0$ change of a single state variable (involves a transition with all external inputs held constant).
    (b) An *externally initiated static-1 hazard* is formed by a $1 \rightarrow 0$ change in a single external input variable (occurs on a holding condition with all y variables constant).
2. A static-0 hazard occurs in product-of-sums (POS) NS forming logic and is a $0 \rightarrow 1 \rightarrow 0$ transient (positive glitch) in an otherwise steady-state (static) logic 0 signal.
    (a) An *internally initiated static-0 hazard* is formed by a $0 \rightarrow 1$ change of a single state variable (involves a transition with all external inputs constant).
    (b) An *externally initiated static-0 hazard* is formed by a $0 \rightarrow 1$ change in a single external input variable (occurs on a holding condition with all y variables constant).

3.3.1   Detection and Elimination of Static Hazards in the NS Forming Logic

There is a certain nomenclature that we use to convey the relatively simple procedures required to detect and eliminate static hazards in two-level logic. The four terms in this nomenclature are as follows:

Coupled variable—a variable that appears uncomplemented in one term of an SOP or POS expression and appears complemented in another term of the same expression.

Coupled term—one of two terms that contains *only one* coupled variable.

Residue—that part of a coupled term that remains after removing the coupled variable.

Hazard cover—the redundant (consensus term) cover required to eliminate the static hazard:

AND—the residues of the two coupled terms to eliminate a static-1 hazard in an SOP expression;

OR—the residues of the two coupled terms to eliminate a static-0 hazard in a POS expression. (See Tinder's book in Endnotes for analyses of static-0 hazards in POS logic.)

In the interest of saving space, we will limit our discussions to static-1 hazards in SOP NS forming logic. We will demonstrate that static hazard identification and elimination can easily be accomplished by using the NS logic functions from K-maps combined with the state diagram from which the K-maps are plotted. In the process, we demonstrate how a normally complicated analysis can be made quite tractable, even routine, for the designer. Note: *The designer must always check the state diagram during a hazard analysis to authenticate the presence of any suspect hazards.*

To illustrate static hazard analysis and elimination, we consider exclusively SOP NS forming logic. Shown in Figure 3.4a is a state diagram with four states, two state variables, two external inputs, and a single conditional (Mealy) output. The NS K-maps in Figure 3.4b apply to both the LPD and nested SR element models. The SR K-maps are generated from the LPD "Y" K-maps by following the $Y \rightarrow SR$ algorithm discussed in Section 2.1. The reader should follow the mapping process beginning with the plotting of the entered variable (EV) K-maps followed by the looping-out process. Once this has been done for the LPD model, attention should next be given to the $Y \rightarrow SR$ transformations. Note that the FSM in Figure 3.4a is devoid of cycles and buffer states.

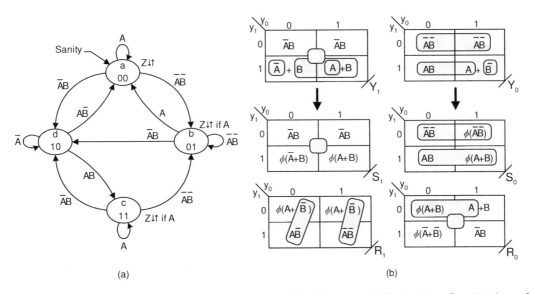

(a)   (b)

FIGURE 3.4: A four-state FSM used to demonstrate identification and elimination of static-1 hazards in SOP NS logic. (a) State diagram with initialization in to the 00 state and a single conditional output, Z. (b) NS K-maps showing optimum SOP cover for both LPD and nested SR element logic designs.

The NS forming logic for the LPD model is obtained form the Y_1 and Y_0 EV K-maps in Figure 3.4b. When looped out as shown, the results are given by Eqs. (3.1). Here, three static-1 hazards are identified all consistent with the state diagram in Figure 3.4a. Their hazard covers (HCs) are indicated by brackets.

$$\begin{cases} Y_1 = \bar{A}B + \overbrace{y_1 B}^{HC} + y_1 y_0 A + y_1 \bar{y}_0 \bar{A} \\ Y_0 = \bar{y}_1 \bar{A}\bar{B} + y_1 AB + y_1 y_0 \bar{B} + y_0 \bar{A}\bar{B} + y_1 y_0 A \end{cases} \tag{3.1}$$

Note in Eqs. (3.1) that there is only one set of coupled terms in the Y_1 expression, an externally initiated static-1 hazard occurring in state 11 during a change $A \to \bar{A}$ while holding on B. The hazard cover for this static hazard is $y_1 y_0 B$, but which is contained in the essential prime implicant (EPI) $y_1 B$. Thus, in this case, an EPI also serves as hazard cover. Although somewhat rare, this is something for which the designer must always check so as to avoid excessive redundant cover.

In contrast, Eqs. (3.1) shows that there are two static-1 hazards present in the expression for Y_0. One is an internally initiated static-1 hazard that takes place on a $11 \to 01$ transition under input conditions $\bar{A}\bar{B}$ (refer to the state diagram in Figure 3.4a). The hazard cover HC_1 is $y_0 \bar{A}\bar{B}$, which permanently eliminates this static-1 hazard; with hazard cover, this static hazard is not possible under any circumstances. The second static-1 hazard is externally initiated occurring in state

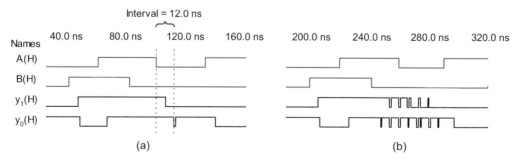

(a)   (b)

FIGURE 3.5: Timing diagram segments showing the effect of the two static-1 hazards in Y_0 of Eqs. (3.1). (a) Without hazard cover HC_1. (b) Combined action of both static-1 hazards in the absence of hazard covers HC_1 and HC_2 resulting in logic instability.

11 under holding condition A when B changes $B \to \bar{B}$. The hazard cover for this hazard is $HC_2 = y_1 y_0 A$, which permanently eliminates the possibility of its formation. The reader will note that both types of hazards result from a $1 \to 0$ change in a coupled variable, a state variable change $y_1 \to \bar{y}_1$ in one case and an external input variable change $B \to \bar{B}$ as the other. The internally initiated static-1 hazard forms as a result of a transition $11 \to 01$ under constant inputs $\bar{A}B$. The externally initiated static-1 hazard forms in state 11 under holding condition A. The actions of these static-1 hazards are shown in Figure 3.5. When both static hazards are active, as shown in Figure 3.5b, unstable conditions occur and malfunction of the FSM most likely results. The interval of 12.0 ns indicated in Figure 3.5a is the time it takes for this hazard (without HC_1) to be activated and is predictably equal to the delay through four gates. For this simulation, the gates and C-elements in the simulator have been set to have a fixed delay of 3.0 ns with inverters set to 1.0 ns.

For comparison purposes, we now run a static-1 hazard analysis on the SR nested element logic. When the SR logic is extracted from the NS EV K-maps in Figure 3.4b, the results are:

$$
\left\{
\begin{array}{ll}
S_1 = \bar{A}B & R_1 = \bar{y}_0 A\bar{B} + y_0 \bar{A}\bar{B} \\
S_0 = \bar{y}_1 \bar{A}\bar{B} + y_1 AB & R_0 = \bar{A}B + \bar{y}_1 A + \bar{y}_1 B
\end{array}
\right\}
\underset{00 \underset{B}{\longleftarrow} 00 \quad \underbrace{}_{HC}}{}
\tag{3.2}
$$

An inspection of the NS forming logic in Eqs. (3.2) indicates that there is a static-1 hazard present in the expression for R_0. The coupled terms $\bar{A}B + \bar{y}_1 A$ in R_0 produce the hazard on a $A \to \bar{A}$ change in state 00 and is eliminated by the consensus term $\bar{y}_1 B$. However, use of C-elements as the memory, serve as filters for hazards developed in the NS forming logic. Thus, there is no need to include the hazard cover term indicated in Eqs. (3.2). This hazard cannot reach the output of the C-element.

Comparing the LPD and nested element NS logic expressions in Eqs. (3.1) and (3.2) indicates a typical difference in the relative complexity of these two design approaches. The total gate/input tally for Eqs. (3.1) is 10/31 (with one shared PI $y_1 y_0 A$), whereas that for Eqs. (3.2) is 10/23 including the shared PI $\bar{A}B$ in the S_1 and R_0 expressions. Not included in these gate/input tallies are the presence of possible inverters, Sanity inputs that would be applied to the presumed NAND logic expressions in Eqs. (3.1), and the inclusion of two C-elements to implement Eqs. (3.2). Inverters can be included in the tallies only if the activation levels of the external inputs are known.

3.3.2   Detection and Elimination of Static Hazards in the Output Forming Logic

Static hazards can occur in the output forming logic as well as in the NS forming logic. However, there is one major difference. A static hazard occurring in the NS forming logic of an asynchronous

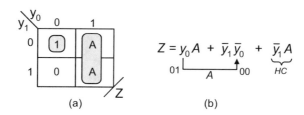

FIGURE 3.6: (a) Output logic K-map plotted from Figure 3.4a. (b) Minimum SOP logic extracted from Figure 3.5a showing an internally initiated static-1 hazard and its cover.

FSM has the potential to cause the FSM to malfunction. A static hazard occurring in the output forming logic can cause a problem only if the transient effect of the hazard can cause a problem in the next stage to which that output is an input. The duration of the transient behavior for all static hazards typically occurs over a period equal to an inverter or gate delay. Shown in Figure 3.6a is the output logic K-map extracted from Figure 3.4a, indicating the presence of an internally initiated static-1 hazard and its cover. This hazard occurs on a $01 \rightarrow 00$ transition under input condition A and is a negative glitch of strength equal to that of an inverter. The timing diagram for this static hazard is given in Figure 3.7 and accompanies Eqs. (3.1). An interval of 16.0 ns is required for production of the static hazard following activation of input A to initiate the transition $01 \rightarrow 00$. This interval is the delay equal to five gate delays plus an inverter.

3.4   DYNAMIC HAZARDS AND FUNCTION HAZARDS

Neither dynamic hazards nor function hazards are static hazards in the sense discussed in Section 3.3. *Dynamic hazards* are multiple glitches in the output such that the output logic levels are different before and after an input change. Dynamic hazards are usually produced in the output of a multilevel circuit due to change in an input for which there are three or more asymmetric paths (delay-wise) of that input to the output. Multilevel AND/OR-type circuits can usually be avoided so that the appearance of dynamic hazards is a nonissue in these circuits. Dynamic hazards that can occur in multilevel exclusive OR (XOR)-type functions are usually unavoidable (see the book by Tinder in Endnotes).

Coupled terms containing two or more coupled input variables can produce *function hazards* if the coupled variables are changed in near proximity to each other. The simplest example of a function hazard is that produced by an XOR function $Y = \bar{A}B + A\bar{B}$ when the inputs are allowed to change in close proximity to each other. The resulting glitch formed in the output Y may or may not cross the switching threshold and this could cause a problem in the next stage to which Y is an input. It is obvious that the potential for function hazard production abounds in asynchronous FSM designs. For example, in Eqs. (3.1) there are two sets of couple terms with two or three coupled variables all of which are capable of producing function hazards. Function hazards in the NS forming logic of asynchronous LPD FSMs can be avoided if their input changes are always minimally separated in time. A function hazard cannot be eliminated by static hazard cover, but use of the nested C-element approach can help to eliminate the effect of function hazards. However, C-elements can go metastable under certain conditions as discussed in Section 3.7.

3.5   OUTPUT RACE GLITCHES—DETECTION AND ELIMINATION

When an asynchronous FSM transitions between two states that involve a change of two or more state variables, a race condition exists. The FSM must transition between an origin state and a destination state via race states as was discussed in Section 3.2. Depending on the outputs involved with respect to the origin and destination states, an ORG may exist. The rules for detecting the presence of an ORG are simply stated as follows:

1. Identify the *origin state* and the *destination state* in a state-to-state transition involving a change of two or more state variables.
2. If the origin and destination states have the same output action relative to a given output, look for the presence of at least one ORG relative to that output which can appear as either a positive or negative glitch.
3. If the origin and destination states have a different output actions relative to a given output, an ORG is not possible relative to that output.

ORGs are timing defects that can be quite disruptive to the proper operation of a given asynchronous FSM. Appearing in an output signal, an ORG can be of strength (time duration) ranging from that for an inverter to several gate delays. In a synchronous FSM, such unwanted transient signals are easily filtered out by the action of the flip-flops and clock signal. However, in an asynchronous FSM there is no suitable means to filter out ORGs so they must be eliminated.

An inspection of the state diagram in Figure 3.4a indicates the presence of an ORG if the FSM transits from origin state 01 to destination state 10 via state 00 under branching conditions $\bar{A}B$. In this case, the origin and destination states have the same output action relative to output Z, that is, neither state can issue an output under branching condition $\bar{A}B$. If the FSM should transit $01 \rightarrow 10$ via race state 00, an ORG will occur. Shown in Figure 3.7 is this ORG that is a positive $0 \rightarrow 1 \rightarrow 0$ glitch of strength equal to the path delay of an inverter. The transition $01 \rightarrow 10$ via race state 11 under branching condition $\bar{A}B$, should it occur, would not produce an ORG because the output action in that state is conditional on an active input A. Setting (Z if A) in state 00 eliminates

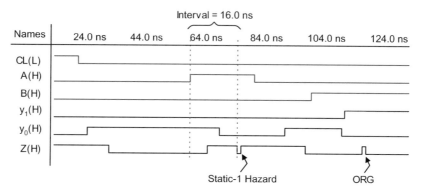

FIGURE 3.7: Timing diagram segment for the output logic given in Figure 3.6 showing a static-1 hazard (exclusive of hazard cover), and an output race glitch (ORG).

the ORG on a $01 \rightarrow 10$ transition. The state diagram given in Figure 1.6a does not have an ORG associated with the transitions $00 \leftrightarrow 11$ based on the three ORG detection rules given at the beginning of this section.

3.6 ESSENTIAL HAZARDS—DETECTION AND ELIMINATION

An active *E-hazard* is a very serious sequential timing defect that can occur as a result of an explicitly located delay in an asynchronous FSM that has at least three states and that is operated in the fundamental mode. Because the FSM can operate properly in the absence of this explicitly located delay, an E-hazard is classified as a *potential* timing defect. As with the other timing defects discussed previously, the fully documented state diagram plays an important role in the identification and elimination of E-hazards. Attempting this process by any other means usually ends up as an exercise in futility.

The development of an E-hazard involves a *race condition* of an initiator input X along two paths to a *race gate* (RG), a direct path and an *indirect path* (IP), as shown in Figure 3.8. If the IP wins the race, an E-hazard is produced. If the direct path wins, as would be the case if no causal delay Δt_E were present, then no E-hazard results. The race conditions can be stated as follows:

Race to First Level RG

E-hazard forms if $\Delta t_E > (\tau_1 + \tau_2) \Rightarrow (y_b$ wins race with initiator $X)$

E-hazard eliminated if $\Delta t_E < (\tau_1 + \tau_2 + \Delta t_{Correction}) \Rightarrow$ (initiator X wins race with $y_b)$

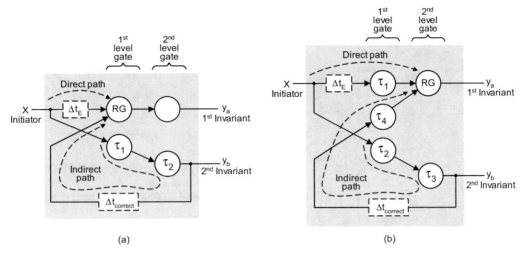

FIGURE 3.8: Illustrations of path delay requirements for E-hazard formation in two-level logic showing causal delays Δt_E, initiator input X, first and second invariants, gate delays τ, race gate (RG), and correction delays to eliminate the E-hazard. (a) First-level race gate indicating direct and indirect delay paths to race gate. (b) Second-level race gate indicating direct and indirect delay paths to race gate.

Race to Second-Level RG
E-hazard forms if $(\Delta t_E + \tau_1) > (\tau_2 + \tau_3 + \tau_4) \Rightarrow (y_b$ wins race with initiator $X)$
E-hazard eliminated if $(\Delta t_E + \tau_1) < (\tau_2 + \tau_3 + \tau_4 + \Delta t_{Correction}) \Rightarrow$ (initiator X wins race with $y_b)$

The delay indicated by $\Delta t_{Correction}$ is a *counteracting delay* placed in a specific feedback path to negate the effect of the inadvertent delay Δt_E, thereby permitting the FSM to function normally.

3.6.1   Minimum Requirements for E-Hazard and D-trio Formation

Shown in Figure 3.9 are the minimum requirements for first-order E-hazard and d-trio formation in NS two-level SOP logic of an asynchronous FSM operated in the fundamental mode. Here, two PS variables y_i and y_j are designated as the first and second invariants, respectively. During transition $a \to b$, present state variable $\bar{y}_i$ remains constant, whereas during transition $b \to c$ state variable y_j remains constant. Note that the second y variable invariant is the first to change, whereas the first y variable invariant is the next one to change. If active, an E-hazard will cause the FSM to erroneously transit $a \to b \to c$ when it should have transitioned $a \to b$, a fatal error in its operation. A *d-trio* is an

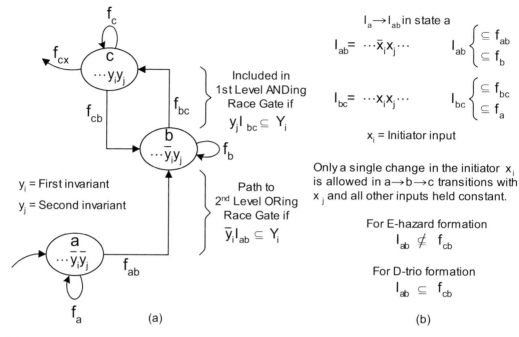

FIGURE 3.9: Minimum requirements for first-order E-hazard and d-trio formation in two-level SOP logic. (a) State diagram segment showing first- and second-level race gate requirements only one of which will be met in the first-invariant function Y_i. (b) Minimum requirements for E-hazard and d-trio formation indicating assumed input conditions for I_{ab} and I_{bc}.

E-hazard that erroneously transits $a \to b \to c \to b$ eventually residing stably in state b where it was intended to go, but by a roundabout path including state c.

With reference to Figure 3.9, the following is a summary of the minimum requirements for activation of E-hazards and d-trios in asynchronous FSMs (here, I denotes an input condition and f is a branching condition):

Summary of Minimum Requirements for E-Hazard Formation

$I_a \to I_{ab} \subseteq f_{ab}$ when $I_{ab} \subseteq f_a$, $I_{bc} \subseteq f_{bc}$, $I_{bc} \subseteq f_a$, and $I_{ab} \not\subset f_{cb}$

Only a single change in the initiator x_i is permitted for $a \to b \to c$ with x_j held fixed.

Summary of Minimum Requirements for D-Trio Formation

Same as for E-hazards except that now $I_{ab} \subseteq f_{cb}$

> ### Summary of IP Requirements for E-Hazard and D-trio Formation as Indicated in Figure 3.9
>
> 1. The IP must be via a gate in the second invariant that cannot contain the RG.
> 2. The IP must *not be inconsistent* with *all* state variables of the initiation state a in Figure 3.9, meaning the state variables $\ldots \bar{y}_i, \bar{y}_j$.
> 3. The IP must *not be inconsistent* with any input held constant during the E-hazard transition, meaning input x_j in Figure 3.9.
> 4. The IP must contain the initiator as either x_i or $\bar{x}_i$.
> 5. The IP must follow a path to the RG that is unobstructed.

Therefore, the IP must be via a gate in Y_j (second invariant), must contain x_i or $\bar{x}_i$, and must not be inconsistent with $\ldots \bar{y}_i \bar{y}_j$ and x_j for SOP (and $\ldots y_i y_j$, and $\bar{x}_j$ for POS). For the sake of brevity, we will not include POS NS forming logic in our discussions.

Note: Always check any static hazard (S-hazard) cover present for possible involvement in the formation of an E-hazard. Static hazard analysis and their elimination must always precede any E-hazard analysis because the effects caused by active static hazards are sometimes similar to those caused by E-hazards.

ANDing race gate test. AND the branching condition for the second transition in a suspect E-hazard path with the second invariant state variable. If the result is contained in the expression for the first invariant, then an ANDing RG exists as depicted in Figure 3.8a. In this case, further analysis for an ORing RG is not necessary. ANDing RGs are more often encountered than ORing RGs.

ORing race gate test. If the test for an ANDing RG is negative, then the analysis must be extended to ORing race gates. An ORing RG means that the race occurs at a second-level gate, which in NAND logic is an ORing operation. To determine whether an ORing RG is present, AND the branching condition for the first transition with the first invariant state variable. If the result is contained in the expression for the first invariant, then an ORing RG exists as depicted in Figure 3.8b. Clearly, the formation of an E-hazard or d-trio by a second-level ORing RG is more unlikely than for a first-level ANDing RG because a larger causal delay Δt_E is required, as indicated by Figure 3.8.

3.6.2 A Simple Example

The detection and elimination of E-hazards is best understood by example, and the use of the fully documented state diagram is the simplest means of accomplishing this. As an example, consider the

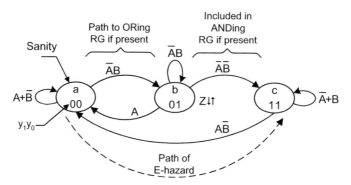

FIGURE 3.10: Simple example of an FSM containing a single E-hazard showing either an ANDing RG or the path to an ORing race gate—one or the other but never both.

FSM presented in Figure 3.10. Here is what we can directly deduce from the state diagram together with Figure 3.9: $\overline{A}\overline{B} \subseteq f_a = A + \overline{B}$, $\overline{A}B \subseteq f_b = \overline{A}B$, and there is only one change in initiator B with ($\overline{A}$ held constant) over the transitions $a \to b \to c$. Furthermore, viewing the $00 \to 01 \to 11$ transitions, it is clear that y_1 is the first invariant and y_0 is the second invariant. Therefore, to initiate the E-hazard, it is required that $\overline{A}\overline{B} \to \overline{A}B$ in state a and that an inadvertent delay Δt_E of sufficient magnitude be explicitly located on the initiator B line to the first invariant Y_1. Also, the IP must be

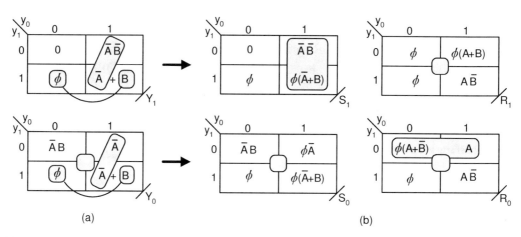

FIGURE 3.11: K-maps derived from Figure 3.10 showing minimum SOP cover. (a) Cover for the LPD design. (b) Cover for the nested SR element logic design.

via a gate in Y_0, must not be inconsistent with $\bar{y}_1, \bar{y}_0, \bar{A}$, and must contain the initiator as either B or $\bar{B}$.

All that now remains is to obtain the NS forming logic for the FSM in Figure 3.10 so as to determine the RG, IP, and the minimum delay (Δt_E) required to activate the E-hazard. Figure 3.11 shows the K-maps for LPD and SR nested element models as extracted from the state diagram in Figure 3.10. Here, we have followed the extended mapping algorithm given in Section 2.1. From these K-maps, we loop out minimum cover for the LPD and nested SR element designs with the results given by Eqs. (3.3) together with the output function Z. Note that an S-hazard exists in NS function Y_1 with hazard cover $y_1 y_0 \bar{A}$ added to eliminate the hazard. No hazards are possible for the SR functions, which is almost always the case.

$$
\begin{aligned}
&\overset{\text{RG}}{Y_1 = \boxed{y_0 \bar{A}\, \bar{B}}} + y_1 B + \underbrace{y_1 y_0 \bar{A}}_{\text{HC}} \quad\longrightarrow\quad \overset{\text{RG}}{S_1 = \boxed{y_0 \bar{A}\, \bar{B}}} \quad R_1 = A\bar{B} \\[2pt]
&\hspace{4.2em}\underset{\bar{A}}{\big|} \\[2pt]
&Y_0 = \boxed{\bar{A}B} + y_0 \bar{A} + y_1 B \quad\longrightarrow\quad S_0 = \boxed{\bar{A}B} \quad R_0 = A\bar{B} + \bar{y}_1 A \\[2pt]
&\hspace{3.3em}\underbrace{\hphantom{\bar{A}B + y_0 \bar{A} + y_1 B}}_{\text{IP}} \qquad\qquad Z = \bar{y}_1 y_0 \qquad \underbrace{\hphantom{\bar{A}B}}_{\text{IP}}
\end{aligned}
\tag{3.3}
$$

$$\underbrace{\hphantom{XXXXXXXXXX}}_{\text{LPD Results}} \qquad\qquad \underbrace{\hphantom{XXXXXXXXXXXXXX}}_{\text{Nested SR Element Results}}$$

The RG and IP can now be determined by viewing the state diagram in Figure 3.10 and by taking into account the minimum requirements for E-hazard formation given in Section 3.6.1. We must ask ourselves the following questions:

1. For an ANDing RG, is $y_0 \bar{A}\bar{B}$ contained in Y_1?   Answer: Yes (see Eq. (3.3))
2. For an ORing RG, is $\bar{y}_1 \bar{A}B$ contained in Y_1?   Answer: No

Therefore, the ANDing RG is $y_0 \bar{A}\bar{B}$ in Y_1 determined by ANDing the branching condition $\bar{A}\bar{B}$ with the second invariant y_0 according to Figures 3.9 and 3.10. Now, the IP must *not be inconsistent* with $\bar{y}_1, \bar{y}_0, \bar{A}$, and must contain B or $\bar{B}$ in Y_0. From Eqs. (3.3), it is clear that the IP requirement is satisfied via gate $\bar{A}B$ in Y_0. RG and IP are both shown in Eqs. (3.3) by shaded areas for both the LPD and nested SR element designs. If we assume that both inputs A and B arrive active high, then the minimum delay Δt_E required to activate the E-hazard in the LPD results is

$$(\Delta t_E + \tau_{\text{Inv}}) > (\tau_{\bar{A}B} + \tau_{Y_0}) \quad \text{or} \quad (\Delta t_E) > (\tau_{\bar{A}B} + \tau_{Y_0} - \tau_{\text{Inv}}) \tag{3.4}$$

Here, τ_{Inv} is included because $B(H)$ must pass through an inverter before it can be an input to RG $y_0 \bar{A}\bar{B}$, where the race occurs. Also, τ_{Y_0} represents the OR function delay required to produce Y_0, as is evident from Eqs. (3.3). Similarly, for a C-element design, the minimum path delay Δt_E required to activate the E-hazard is from Eqs. (3.3) given by

$$(\Delta t_E + \tau_{Inv}) > (\tau_{\overline{AB}} + \tau_{C\text{-}element}) \text{ or } (\Delta t_E) > (\tau_{\overline{AB}} + \tau_{C\text{-}element} - \tau_{Inv}) \qquad (3.5)$$

where $\tau_{C\text{-}element}$ is the propagation delay through the C-element.

The logic circuit for the C-element design, as required by Eqs. (3.3), is presented in Figure 3.12. Here, we include the direct and IPs, RG, indirect path gate, and the location of the inadvertent causal delay Δt_E. The markings and labels on this circuit are consistent with Figure 3.8. We have used the *wireless connection* feature as recommended for use with the simulator described in the Preface. Also included in Figure 3.12 is the output read from the state diagram as

$$Z = \overline{y}_1 y_0 \qquad (3.6)$$

This output is of interest to us at this point only if the E-hazard becomes active. An E-hazard transition $a \rightarrow b \rightarrow c$ would cause the output Z to glitch over a period equal to two gate delays as indicated in the timing diagram of Figure 3.13.

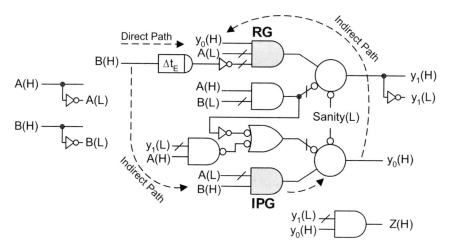

FIGURE 3.12: C-element circuit using the wireless connection feature for the SR logic in Eqs. (3.3) showing the RG, IPG, and the race between the direct path of the initiator B and the indirect path of the second invariant y_0.

In our simulations up to this point, we have, for simplicity, assigned a propagation delay of 3.0 ns to all gates and C-elements, and a propagation delay of 1.0 ns to each inverter. (The simulator does permit a wide variation of assigned gate, C-element, and inverter delays.) With this in mind, the minimum path delay to activate the E-hazard for either Eqs. (3.4) or Eqs. (3.5) is

$$\Delta t_E > (3 + 3 - 1) = 5 \text{ ns.} \qquad (3.7)$$

This result is easily tested by simulation. To do this, we take the C-element design for simplicity. Shown in Figure 3.13 are the simulations for two settings of Δt_E, one exactly at 5.0 ns for normal operation, and the other at 5.1 ns causing the formation of the E-hazard. Note that, in the latter case, the FSM transits incorrectly 00 → 11 via state 01 spending two gate delays ($\tau_{AND} + \tau_{C\text{-element}} = 6.0$ ns) in that race state. What is happening here is that the delay Δt_E, of sufficient magnitude, allows the second invariant to win the race with the initiator B at the RG, thereby permitting the FSM to cycle from state 00 to 11 via state 01. In effect, the FSM tries to execute the proper state change required by $\overline{A}\overline{B} \rightarrow \overline{A}B$ but because y_0 wins the race with initiator B, the FSM still senses $\overline{A}\overline{B}$ and transits on to state 11, after which $\overline{A}B$ becomes valid. If Δt_E is not of sufficient magnitude to activate the E-hazard, normal operation will occur.

An active E-hazard can effect an output involved in the E-hazard path. This is illustrated in Figure 3.13, where an output glitch of 6 ns is produced as a result of the E-hazard path 00 → 01 → 11, where the FSM spends 6 ns in state 01. Of course, if the E-hazard is absent, the E-hazard OP

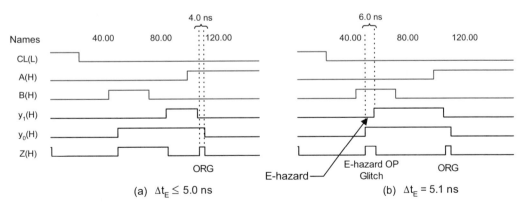

(a) $\Delta t_E \leq 5.0$ ns   (b) $\Delta t_E = 5.1$ ns

FIGURE 3.13: Timing diagrams for the C-element design of Eqs. (3.3) showing the effect of an active E-hazard. (a) Normal operation with $\Delta t_E \leq 5.0$ ns. (b) FSM malfunction due to an E-hazard formation when $\Delta t_E > 5.0$ ns.

glitch disappears. Note that with or without the activation of an E-hazard, an ORG is produced during the 11 → 00 transition via race state 01. This is a relatively large ORG of 4-ns duration. However, this ORG can be easily eliminated by making the output Z conditional on $\bar{A}B$. Now, the branching 11 → 00 caused by the branching condition $A\bar{B}$ cannot activate Z during a race path via state 01. The large ORG demonstrates that such timing defects can be very disruptive should the output be presented to a next stage that can be affected by such transient output signals.

D-trio formation. So far, we have not demonstrated the formation of d-trios mainly because an active d-trio may or may not be disruptive to the operation of the FSM. For completeness, we draw the readers' attention to the FSM in Figure 3.4. There are three d-trios identified from this state diagram. They are $a \to d \to c \to d$, $d \to a \to b \to a$, and $c \to b \to a \to b$. We will take the first one 00 → 10 → 11 → 10 for demonstration purposes. Here is what we can deduce from the state diagram together with Eqs. (3.1): The initiating transition is $AB \to \bar{A}B$, the initiator is A with B held constant. The ANDing RG is y_1AB in Y_0, y_0 = first invariant and y_1 = second invariant. Δt_E must exist on the A line to Y_0. The IP must not be inconsistent with $\bar{y}_1, \bar{y}_0, B$; must contain A or $\bar{A}$; and must exist via a gate in Y_1. Therefore, the IP must be $\bar{A}B$. From this information we calculate the minimum causal delay to be $\Delta t_E > (\tau_{\text{Inv}} + \tau_{Y_1})$. If y_1 wins the race with initiator A at the RG y_1AB, then the FSM will transit to state 11 where, after a short while, it will sense $\bar{A}B$ and transit back to state 10. Given the delays we have assigned to the simulator used in this text, $\Delta t_E > (\tau_{\text{Inv}} + \tau_{Y_1}) = 4$ ns. To show an understanding of the analysis process, the reader should prove that the d-trio $c \to b \to a \to b$ requires an ORing RG if $A\bar{B} \to \bar{A}\bar{B}$ occurs in state 11, and that the path to the ORing RG is via $y_0\bar{A}\bar{B}$ in Y_0, which is a hazard cover. Now find the IP.

Elimination of E-hazards. Inadvertent delays Δt_E occur for various reasons; the most common are due to manufacturing errors such as those made at the foundry. For this reason, the designer cannot disregard the possibility that an E-hazard of sufficient magnitude may appear at a specific location. So what can be done to eliminate the possible formation of the E-hazards that may exist in a given FSM? The answer is simple and is depicted in Figure 3.8. All that is necessary is to place a counteracting delay $\Delta t_{\text{Correct}}$ in the feedback of the second invariant state variable. A conservative value for this counteracting delay would be equal to Δt_E as determined from E-hazard analysis. But what if E-hazards and d-trios abound in a complex state diagram? Some designers would recommend that the "shotgun" approach be used—that is, the placing of counteracting delays on *all* feedback lines. The problem with this approach is that these delays placed on all feedback lines may significantly slow down the operation of the asynchronous FSM. So, whenever it is possible and justifiable, the designer should take the time to run the required analyses to determine where the E-hazards exist and assign reasonable values of $\Delta t_{\text{Correct}}$ to be placed in certain specific feedback lines. If only one or two E-hazards exist in a relatively complex FSM, it would be a wise move to treat them separately. Also, it is known that some E-hazards require such large Δt_E that nothing may need to be done to prevent their becoming active. However, there is one caveat that should be mentioned here. Modern designs now use gates (or their metal-oxide semiconductor equivalent) that are extremely fast, having very small propagation delays. Line delays caused by transmission line effects, for example, can cause E-hazards in such circuits. Thus, the designer of such modern circuits would be well advised to consider gate propagation delays relative to circuit or chip layout to make an intelligent decision as to what, if anything, must be done regarding possible E-hazard formation. If counteracting delays $\Delta t_{Correct}$ are necessary in y-variable feedback lines, use of an even number of inverters is usually the best choice. Alternatively, inertial delay circuits can be used, but at an additional cost in hardware together with an unnecessary increase in response time. An inertial delay will usually consist of capacitors, diodes, resistors, and a rendevous module (see Tinder's text in Endnotes).

This brings us to reiterate some important points. E-hazards are "sequential" hazards that are strictly a function of the sequential behavior of the asynchronous FSM as evidenced from an inspection of the state diagram representation. FSMs of three or more states are subject to possible E-hazard formation, but only in state machines operated in the fundamental mode. In such machines, they are nearly always regarded as potential timing defects that can be activated only when delays exceeding a minimum value are placed in specific locations in the circuits. If active, E-hazards are guaranteed to cause the malfunction of the asynchronous FSM. The only reliable means of eliminating an E-hazard is to place a counteracting delay in the feedback of the second invariant state variable as discussed in Section 3.6.

3.7   METASTABLE CONDITIONS IN C-ELEMENTS

We have purposely left the reader with the impression that FSMs designed with C-elements lead to quasi-Muller designs, implying that the C-elements are free of metastable conditions arising from some specific input condition. This is only partially true. It is true that static-1 hazards originating in the NS forming logic will normally not pass through the C-elements because the C-elements operate outside of the fundamental mode. This is an important feature of C-elements assumed to have inputs from SOP NS forming logic. However, if an input to a C-element is withdrawn before the "week (keeper)" feedback inverter fully responds to the last input change, a static hazard can be formed and a metastable condition can result. This fact as been proven by SPICE simulations. Modern high-speed CMOS inverters have a very small propagation delay time, perhaps less that a picosecond. Here, we are talking about pulse widths that may or may not cross the switching threshold. So, even if the probability of a metastable occurrence in C-elements is extremely small, it can still happen. Thus, we conclude the obvious: *There is no logic device that can be guaranteed free of internal conditions leading to metastability.* Whether the use of input conditioning arbiters can be justified in C-element-based FSMs is a matter that the designer must weigh. It is doubtful that any input conditioning arbiter module can properly deal with the "keeper inverter" issue. However, the use of a properly designed arbiter can detect a metastable condition and prevent that metastability from being passed on to the next stage. Chapter 11 discusses bus and handshake arbiters that effectively deal with the metastability problem.

CHAPTER 4

Design of Single Transition Time Machines

We now describe a class of asynchronous finite state machines (FSMs) whose transition times are the fastest possible and that avoid all race-associated timing defects, namely, critical races and output race glitches (ORGs). This class of fundamental mode FSMs are commonly called *single transition time (STT) machines*. Here, state code assignments must be found that will eliminate critical races and ORGs while at the same time producing next state (NS) functions that represent the fastest transition times possible. The means by which this can be accomplished is called the *array algebraic approach* to state machine design. This approach lends itself nicely to computer-aided design (CAD) all without the use of state diagrams or K-maps. Instead, use will be made of the *state table* and partitioning methods as discussed in the following sections.

4.1 THE ARRAY ALGEBRAIC APPROACH

To assist in the understanding of the array algebraic approach to asynchronous FSM design, we outline the procedure as follows:

1. By using a state diagram, construct the *state table* free of cycles and buffer states both of which are strictly forbidden. Use state identifiers (*a*, *b*, *c*, ...) in the state table as was done in Figure 1.6b and make certain that the *sum rule* holds for all states. Violation of the sum rule can cause critical races. If any two rows in the state table are identical with respect to the state identifiers, *merge* these two rows into one and rename the state identifiers accordingly.

2. Identify the state that is to be initialized and assume that it will be an all-zero state (...000) or an all-one state (...111) following the initialization procedure given in Section 2.2. Although these assumptions are not mandatory, they do simplify considerably the initialization process.

3. Partition the state transitions into sets that eliminate critical races and ORGs. This is done by the use of π (partial)-partitions gathered from the input columns of the state table by applying an *extension* of the "into rule" stated as follows:

> The **into rule**: Make logically adjacent assignments to present states that branch "into" a common state, provided that their input conditions are the same.

The state identifier for the initiation state, together with all the other state identifiers associated with that initialization state, must be positioned on the left side of the π-partitions (separated by a comma). In doing this, a valid STT state code assignment can be obtained by following the remaining steps of this procedure.

4. Collect the π-partitions that include *all* state identifiers into τ (total)-partitions such that each τ-partition begins with the state identifier for the initialization state and all other associated state identifiers on the left side of the partition (indicated by a comma).

5. Find the minimum set of τ-partitions that "cover" all π-partitions. The resulting number of τ-partitions is equal to the minimum number of state variables for the FSM. If there is more than one minimum set of τ-partitions, any one of the minimum sets will yield an optimum or near optimum STT design—there is usually little difference in their use.

6. Select a valid state code assignment for the FSM from a minimum set of τ-partitions choosing the initialization state to be either an all-zero state (…000) or an all-one state (…111), not a mixture. Note that for FSMs lacking *cross branching*, the partitioning methods default to a unit distance (Hamming distance of one) coding of states as for the corrected FSM in Figure 3.3.

4.2   DESIGN EXAMPLE USING C-ELEMENTS

As an example, consider the FSM presented in Figure 4.1a, which contains two *cycle paths* that are eliminated in Figure 4.1b. In eliminating the cycle paths in Figure 4.1a, it is clear that the transition $c \to a$ under branching condition $\bar{A}\bar{B}$ is meant to be $c \to b$, and the transition $a \to b$ under branching condition $\bar{A}B$ is meant to be $a \to c$.

The state table representing Figure 4.1b is given in Figure 4.1c, and its significance follows the description of the state table in Figure 1.6b. The state identifiers that are encircled indicate the holding conditions that satisfy the state diagram and the stability criteria given by Eqs. (1.1) and (1.2). From the state table we can obtain the π-partitions and finally the τ-partitions from which the state code assignments derive. The π_1-, π_3-, and π_2-partitions are given below from columns $I_1 = \bar{A}B$, $I_3 = AB$, and $I_2 = A\bar{B}$, respectively.

$$\pi_1 = abc, d = \tau_1 \text{ from } I_1; \quad \pi_3 = ab, cd = \tau_3 \text{ from } I_3; \quad \pi_2 = ad, bc = \tau_2 \text{ from } I_2$$

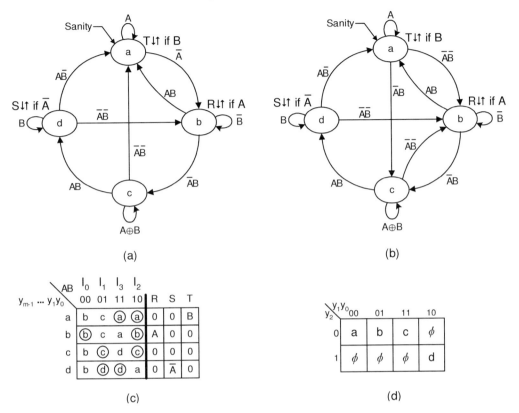

FIGURE 4.1: (a) State diagram for an FSM having two cycle paths. (b) The state diagram in (a) with the cycle paths removed. (c) State table for the state diagram in (b). (d) State assignment map obtained from the state matrix, S.

Note that τ_2 combines π-partitions $\pi_4 = ad, b$ and $\pi_5 = ad, c$, the numbering system being a matter of choice. Because there are three τ-partitions, there must be a minimum set of three state variables with a chosen as the initiation state. If we choose to initialize into the 000 state, a valid *state assignment matrix, S* , is given in Eqs. (4.1), which is used to plot the state assignment map in Figure 4.1d.

$$
S = \begin{array}{c} \\ a \\ b \\ c \\ d \end{array}
\begin{bmatrix}
\tau_1 & \tau_3 & \tau_2 \\
0 & 0 & 0 \\
0 & 0 & 1 \\
0 & 1 & 1 \\
1 & 1 & 0
\end{bmatrix}
\qquad
D = \begin{array}{c} \\ a \\ b \\ c \\ d \end{array}
\begin{bmatrix}
I_0 & I_1 & I_3 & I_2 \\
0 & 0 & ab & ad \\
1 & 0 & 0 & b \\
0 & abc & 0 & c \\
0 & d & cd & 0
\end{bmatrix}
\qquad (4.1)
$$

The column arrangement in the S matrix is a matter of choice because there are $3! = 6$ ways to permute the three τ-partitions and hence 6 possible and valid state code assignments. We have chosen the state assignment given in S specifically for later comparison with the computer automated design of this FSM given in Section 4.4. If initializing into $a = 111$ state is an alternative, then there are $2 \times 3! = 12$ possible and valid state code assignments. Generally, there are $n!$ valid state code assignments for n state variables when the initialization state is chosen to be the all zeros state, or $2n!$ valid state code assignments if initialization into the all ones state is an alternative.

The *destination matrix* D is also given in Eqs. (4.1). It is constructed from either the state table or state diagram by identifying those states whose destination is to a specific present state identifier under a given branching condition. For example, under branching condition $I_0 = \bar{A}\bar{B}$, state b exclusively branches to itself (a holding condition). Or, under branching condition $I_1 = \bar{A}B$ states a, b, and c branch to state c, whereas state d branches to itself as a holding condition. The "1" in the I_0 column for D results from the state association $abcd = 1$.

Now, with the S and D matrices known, we can find the *function matrix*, F_{NS} given by

$$
F_{\text{NS}} = \tilde{S}D =
\begin{bmatrix}
0 & 0 & 0 & 1 \\
0 & 0 & 1 & 1 \\
0 & 1 & 1 & 0
\end{bmatrix}
\begin{bmatrix}
0 & 0 & ab & ad \\
1 & 0 & 0 & b \\
0 & abc & 0 & c \\
0 & d & cd & 0
\end{bmatrix}
=
\begin{bmatrix}
0 & d & cd & 0 \\
0 & 1 & cd & c \\
1 & abc & 0 & bc
\end{bmatrix}
$$

$$
=
\begin{bmatrix}
0 & y_2 & y_1 & 0 \\
0 & 1 & y_1 & \bar{y}_2 y_1 \\
1 & (\bar{y}_1 + y_0) & 0 & y_0
\end{bmatrix}
$$

(4.2)

Here, $\tilde{S}$ is the transpose of S, the matrix multiplication result $abcd = 1$, and the y-variable matrix on the right side is one of a few alternative forms obtained from the state assignment map in Figure 4.1d. Notice that Figure 4.1d permits $\bar{y}_2 y_1$ or $y_1 y_0$. Although the y-variable matrix in Eqs. (4.2) is handpicked for the state assignment chosen, the reader must note that for complex asynchronous state machines it may be necessary to use a logic minimizer to obtain an optimum y-variable matrix from the state assignment map. This is done within the CAD program called ADAM (Automated Design of Asynchronous Machines), described in Section 4.4.

The NS functions can now be found from the following conformable matrix multiplication, where it will be recalled that $I_0 = \bar{A}\bar{B}$, $I_1 = \bar{A}B$, $I_3 = AB$, and $I_2 = A\bar{B}$:

$$
F_{\text{NS}}I =
\begin{bmatrix}
0 & y_2 & y_1 & 0 \\
0 & 1 & y_1 & \bar{y}_2 y_1 \\
1 & \bar{y}_1 + y_0 & 0 & y_0
\end{bmatrix}
\begin{bmatrix}
I_0 \\
I_1 \\
I_3 \\
I_2
\end{bmatrix}
=
\begin{bmatrix}
Y_2 \\
Y_1 \\
Y_0
\end{bmatrix}
=
\begin{bmatrix}
y_2 \bar{A}B + y_1 AB \\
\bar{A}B + y_1 AB + \bar{y}_2 y_1 A\bar{B} \\
\bar{A}\bar{B} + (\bar{y}_1 + y_0)\bar{A}B + y_0 A\bar{B}
\end{bmatrix}
$$

(4.3)

After simplification by applying the factoring and absorptive laws (see Appendix A.2), there results the final two-level expressions for Y_2, Y_1, and Y_0 given by

$$Y_2 = y_2\overline{A}B + y_1AB + \underset{HC}{y_2y_1B}$$
$$Y_1 = \overline{A}B + y_1B + \overline{y}_2y_1A \qquad (4.4)$$
$$Y_0 = \overline{A}\overline{B} + \overline{y}_1\overline{A} + y_0\overline{A} + y_0\overline{B}$$

where HC below the term y_2,y_0B indicates that term is a static 1-hazard cover for the coupled terms $y_2\overline{A}B$, y_1AB. Without the hazard cover, this hazard would cause a 1–0–1 glitch in Y_2 when $A \to \overline{A}$ in state 110 = d under holding condition B. The y-variable matrix in Eqs. (4.2) interpreted from the state assignment map in Figure 4.1d eliminates all other S-hazards. (See Section 3.3.1 for a review of S-hazards in the NS forming logic.)

Following a similar procedure, the three outputs are obtained as follows:

$$F_R = \tilde{R}DI = \begin{bmatrix} 0 & A & 0 & 0 \end{bmatrix} DI = \begin{bmatrix} 0 & A & 0 & A\overline{y}_1y_0 \end{bmatrix} I \quad R = \overline{y}_1y_0\overline{A}B$$
$$F_S = \tilde{S}DI = \begin{bmatrix} 0 & 0 & 0 & \overline{A} \end{bmatrix} DI = \begin{bmatrix} 0 & \overline{A}y_2 & \overline{A}y_1 & 0 \end{bmatrix} I \quad S = y_2\overline{A}B \qquad (4.5)$$
$$F_T = \tilde{T}DI = \begin{bmatrix} B & 0 & 0 & 0 \end{bmatrix} DI = \begin{bmatrix} 0 & 0 & B\overline{y}_1 & B\overline{y}_0 \end{bmatrix} I \quad T = \overline{y}_1AB$$

Here, I represents the input column matrix as in Eqs. (4.3). Notice that the term $y_2\overline{A}B$ for S is a shared PI with the first term in Y_2. Such shared PIs are common in the array algebraic method. This brings the total NS and output gate/input tally for the LPD results to 15/40 exclusive of inverters.

Now that the LPD NS forming logic has been found by using the array algebraic approach, we will use $Y \to SR$ K-map conversion to obtain the set–reset (SR) logic required for C-element or SR basic cell design of this FSM. To do this, we will plot the NS logic in Eqs. (4.4) in third-order entered variable (EV) K-maps so that the essence of the array algebraic approach is maintained. Then, using these K-maps, we will convert to SR EV K-maps following the procedure outlined and demonstrated in Section 2.1. The K-map $Y \to SR$ conversions are shown in Figure 4.2, where optimum SR logic is indicated by shaded loops. The reader should follow the plotting and loop-out processes closely so that the procedures are fully understood. (Refer to Appendix A.3 for a review of EV K-map minimization.) The SR results are given below.

$$
\begin{array}{ll}
S_2 = y_1AB & R_2 = \overline{B} + \overline{y}_1A \\
S_1 = \overline{A}B & R_1 = \overline{A}\overline{B} + y_2\overline{B} \\
S_0 = \overline{A}\overline{B} + \overline{y}_1\overline{A} & R_0 = AB
\end{array} \qquad (4.6)
$$

Notice that there is one shared PI, $\overline{A}\overline{B}$, bringing the total NS and output gate/input tally to 13/31 exclusive of C-elements and inverters. When the three C-elements are included, the total NS and

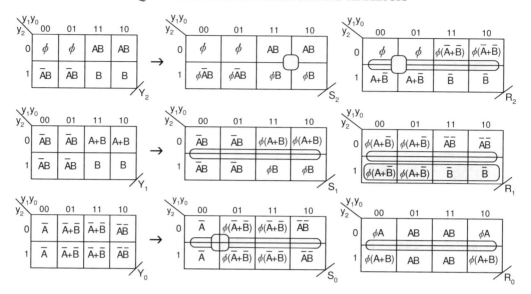

FIGURE 4.2: The NS logic LPD-to-SR K-map conversion showing optimum SR cover for use with Muller C-elements or SR basic cells.

output gate/input tally becomes 16/37, again exclusive of inverters. Inverters are not tallied unless their input activation levels are known. Note also that the sanity circuit initialization input to the C-elements is not indicated in Eqs. (4.6). That is because the initialization input overrides the NS forming logic permitting normal operation of the FSM beginning from the initialization state (state $a = 000$ in this case) only after the sanity input has been "turned off."

Shown in Figure 4.3 are two implementations of the asynchronous FSM in Figure 4.1b. These are the lumped path delay (LPD) model by using Eqs. (4.4), and the SR design with C-elements by using Eqs. (4.6). In both cases, inputs A and B are assumed to arrive active high, and the sanity circuit input is assumed to be that of Figure 2.5a with input requirements illustrated in Figure 2.7a and 2.7c. Notice that in the case of the LPD model in Figure 4.3a, the Sanity(L) inputs must be applied via the ANDing operations to the NS logic. In contrast, the C-element implementation in Figure 4.3b allows the Sanity(L) inputs to be applied directly to the C-elements. However, in either case, initialization is an override forcing the FSM into the $a = 000$ state ready for the FSM to operate normally once the Sanity(L) input changes $1(L) \rightarrow 0(L)$. Notice that the *wireless connection feature* is used in Figure 4.3 to simplify the circuit and thereby minimize the chance for error, as will be the practice throughout this text. This is the same schematic capture feature that is recommended for use with the logic simulator described in the Preface.

Simulation of the C-element implementation in Figure 4.3b is shown in Figure 4.4 together with the outputs in Figure 4.3d. The reader should verify the validity of this timing diagram by comparing it with the state diagram given in Figure 4.1b. One advantage of the C-element design is that it minimizes the effects of *function hazard* formation in the NS logic when inputs change in close proximity to each other. This advantage results from the fact that C-elements operate

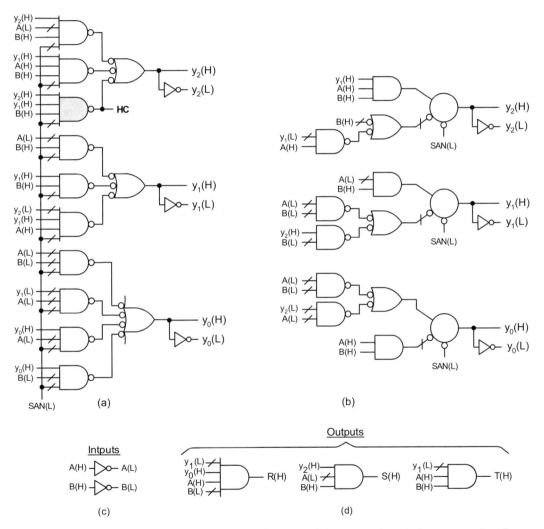

FIGURE 4.3: Implementation of the FSM in Figure 4.1b by using the wireless connection feature. (a) The LPD model by using Eqs. (4.4) including hazard cover indicated by the shaded gate. (b) The SR design with C-elements by using Eqs. (4.6) with no hazard cover needed. (c) Inputs. (d) Output logic.

outside the fundamental mode. However, the reader must be warned that if the inputs to an output function change in close proximity to each other, a function hazard can occur in that output signal. Remember that a function hazard actually represents a "proper response" to inputs that are allowed to change close to each other. The problem arises when the inputs change so close to together that an output spike occurs that might or might not cross the switching threshold. This potential problem is eliminated by requiring that input changes to the output logic be minimally separated in time. However, internally initiated function hazards (by y variables) can occur and are difficult, if not impossible, to eliminate. Static hazards are rare in the output logic functions that derive from the array algebraic approach, regardless of the model used.

A simulation of the LPD circuit in Figure 4.3a will differ only slightly from that in Figure 4.4 because of the difference in throughput delays of the two implementations. The C-element design is basically a three-level design, whereas that for the LPD circuit is a two level design—referring to levels of path delay. This difference is somewhat offset by the fact that the average number of gate inputs is a little more than 3 for the LPD design compared to about 2 for the C-element design. Remember that gate propagation delay increases with number of inputs. Thus, we conclude that the two designs are expected to have about the same power/delay product.

The LPD and SR designs of the FSM in Figure 4.1b, without the use of the array algebraic approach, can be done as in Figures 3.10 and 3.11. By using a logic minimizer (such as BOOZER, described in the Preface), simplified results can be obtained. However, in doing this, ORGs and critical races are now possible. The special state assignment partitioning methods used in the array algebraic approach guarantees that these timing defects will not exist in the resulting logic. As a

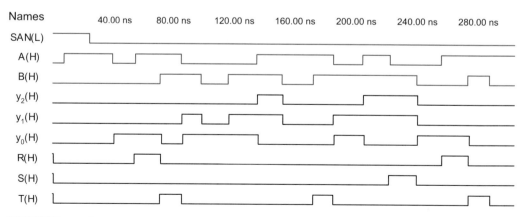

FIGURE 4.4: Simulation of the C-element logic circuit in Figure 4.3b combined with the outputs in Figure 4.3d.

rule, the presence of cross branching will allow the presence of ORGs and/or critical races unless sufficient numbers of state variables are added to remove any race conditions.

4.3   ESSENTIAL HAZARD ANALYSIS IN STT FSMS

It is appropriate at this time to conduct an E-hazard analysis of the FSM in Figure 4.1b. This can be done without the need for a logic circuit simply by using the state diagram in Figure 4.1b and the NS logic given in Eqs. (4.4) or Eqs. (4.6), keeping in mind the information provided in Sections 3.6.1 and 3.6.2. Also, it must be remembered that an E-hazard is only a potential timing defect caused by an unintended delay occurring at some specific place in the asynchronous circuit.

E-Hazard #1. The required path is $a \rightarrow c \rightarrow d$ $(000 \rightarrow 011 \rightarrow 110)$ with delay Δt_{E1} on A (the initiator) to y_2 (first invariant) under input conditions $AB \rightarrow \bar{A}B$.

Question: Is an ANDing race gate (RG) = $y_1 AB$ present in Y_2 or S_2? Answer: Yes.

Question: Is there an indirect path (IP) not inconsistent with the origin state $(\bar{y}_2, \bar{y}_1, \bar{y}_0)$, constant input B, and that must contain *(A or $\bar{A}$)* in the second invariant y_1? Answer: Yes, via $\bar{A}B$ in Y_1 or S_1.

Therefore, an E-hazard will be activated if the second invariant y_1 wins the race with A at the RG = $y_1 AB$ due to the delay $\Delta t_{E1} > (2\tau_P + \tau_{Inv})$ on the A line to the first invariant y_2. Here, τ_{Inv} is required for A to reach gate $\bar{A}B$ via an inverter, assuming A arrives active high, and the $2\tau_P$ term contains two delays $2\tau_P = \tau_{\bar{A}B} + \tau_{ORing}$ for the LPD circuit or $2\tau_P = \tau_{\bar{A}B} + \tau_{C\text{-element}}$ for the C-element circuit. The reader should use either circuit in Figure 4.3 to trace the direct and indirect paths for the formation of this E-hazard. A logic simulation of this E-hazard will show that $\Delta t_{E1} > (2\tau_P + \tau_{Inv})$ is precisely correct regardless of what propagation delays are chosen for the circuit elements. This E-hazard can be safely eliminated by placing a counteracting delay $\Delta t_{Correct} = \Delta t_{E1}$ on the y_1 feedback line to RG. This delay is obviously quite conservative because all that is necessary is to ensure that $\Delta t_{E1} < (2\tau_P + \tau_{Inv} + \Delta t_{Correct})$.

E-Hazard #2. The required path is $c \rightarrow d \rightarrow a$ $(011 \rightarrow 110 \rightarrow 000)$ with delay Δt_{E2} on B (the initiator) to y_1 under input conditions $A\bar{B} \rightarrow AB$. Here, y_1 is the first invariant and y_0 is the second invariant as deduced from $011 \rightarrow 110 \rightarrow 000$. The following pertains to the LPD mode only.

Question: Is an ANDing race gate RG = $\bar{y}_0 A\bar{B}$ present in Y_1? Answer: No.

Question: Is an ORing RG via path $y_1 AB$ present in Y_1? Answer: Yes, via $y_1 B$ in Y_1.

Question: Is there an IP not inconsistent with the origin state $(\bar{y}_2, y_1, y_0)$, constant input A, and that must contain $(B \text{ or } \bar{B})$ in Y_0? Answer: Yes, via $y_0 \bar{B}$ in Y_0. However, there is no gate in Y_1 containing the first invariant y_0 or $\bar{y}_0$ that allows completion of the IP path. Therefore, E-hazard #2 cannot exist under the existing NS functions.

Had we interpreted $c = y_1 y_0$ instead of $\bar{y}_2 y_1$ in Eqs. (4.2) and (4.3), the result for Y_1 would be the OPIs $Y_1 = \bar{A}B + y_1 B + y_1 y_0 A$ for the LPD model and $S_1 = \bar{A}B$ and $R_1 = \bar{A}\bar{B} + \bar{y}_0\bar{B}$ for the SR model. Note that no valid S-hazard exists for the coupled terms $\bar{A}B$ and $y_1 y_0 A$. But now the term $y_1 y_0 A$ in Y_1 provides entrance to Y_1 via its y_0 input as required to complete the IP. Note that the new interpretation $c = y_1 y_0$ does permit an ANDing RG in R_1 and an IP via R_0. Therefore, we now conclude that the minimum path delay required for E-hazard #2 formation is $(\Delta t_{E2} + \tau_P) > (\tau_{Inv} + 3\tau_P)$, assuming that input B arrives active high. Here, τ_P on the left side represents the path delay through $y_1 B$, whereas $3\tau_P$ on the right side represents path delays through gates $y_0 \bar{B}$, Y_0 (ORing), and $y_1 y_0 A$ in Y_1 in that order. Thus, if y_0 wins the race with A to the three-input NAND gate performing the ORing operation for y_1, the E-hazard will be activated. This E-hazard can be prevented by placing a counteracting delay $\Delta t_{Correct}$ on the y_0 feedback line such that $(\Delta t_{E2} + \tau_P) < (\tau_{Inv} + 3\tau_P + \Delta t_{Correct})$.

With the Y_1 changes given above, the reader should mentally trace through the direct and indirect paths for this E-hazard in the LPD circuit of Figure 4.3. Note that E-hazard #2 is not possible for the SR model because it does not satisfy all requirements for E-hazard formation via an ORing RG.

4.4 COMPUTER AIDED STT FSM DESIGN

ADAM (for Automated Design of Asynchronous Machines) is a unique, versatile, and powerful CAD tool that permits the automated design of complex STT asynchronous FSMs. ADAM designs are free of all timing defects in the NS logic with the exception of function hazards, and static hazards can exist in the output logic. Depending on the command options used, as described in the *Readme.doc* accompanying the software, ADAM can be used to design LPD model circuits, or SR model circuits that use complementary C-elements or SR basic cells as the memory elements. ADAM can also be used to directly program programmable logic arrays (PLAs) in the Berkeley format. To run ADAM, it must be located in the C:\ADAM directory together with two .txt files described below.

Shown in Table 4.1 is the STT *input file* Fig41c (no punctuation) constructed by using a text editor such as Notepad as required by ADAM. For comparison, we will use this file with ADAM to design the FSM represented by the state table in Figure 4.1c. The remarks given on the right side of the table are self-explanatory. However, it is important that the reader remember that *the inputs in any state table suitable for use with ADAM must be unfolded in binary* not in gray code as in Figure 4.1c. Thus, the last two columns in Figure 4.1c must be interchanged for use with the ADAM STT input file Fig41c. The reader should compare the state table in Figure 4.1c with that in Table 4.1 for proof that they are identical—remember that the AB inputs in the two tables are unfolded differently, one in 2-bit Gray code and the other in 2-bit binary. This state table is free of cycle paths and buffer states as required for any STT design that uses the array algebraic approach or ADAM.

Once the input file for ADAM has been constructed, it must be called from a user-generated *batch file*, which we will call Fig41c.bat (again, no punctuation). This batch file, given in Table 4.1, is also constructed by using a .txt editor such as Notepad. Notice that it has seven parts to it. This is

TABLE 4.1: CAD design of the FSM in Fig. 4.1(c) by using software called ADAM. (a) STT input file, Fig41c (b) Remarks on the input file. (c) Batch file Fig41c.bat as required by ADAM for an SR design of Fig. 4.1(c).

(a) ADAM Input File	(b) Remarks
4	Number of rows in the state table equals the number of states
5	Columns in the state table including the present state decimal column
2	Number of input variables to the machine
3	Number of output variables for the machine
A B #	Names of input variables separated by a space and terminated by #
R S T #	Names of output variables separated by a space and terminated by #
N	User to supply the state code assignment? Y or N
	—(Space required here)—
0 1 2 0 0	
1 1 2 1 0	State table in Fig. 4.1(c) with inputs AB unfolded in *binary* not gray code.
2 1 2 2 3	In the first column are the numerical state identifiers ($a = 0, b = 1, c = 2$, etc.)
3 1 3 0 3	Place one space between columns; Spaces between rows not required.
	—(Space required here)—
0 0 0 0	
0 0 1 1	Table for output R from Fig. 4.1(c) with inputs AB unfolded in *binary*.
0 0 0 0	Numerical state identifier column not entered but assumed.
0 0 0 0	Place one space between columns; Spaces between rows not required.
	—(Space required here)—
0 0 0 0	
0 0 0 0	Table for output S from Fig. 4.1(c) with inputs AB unfolded in *binary*.
0 0 0 0	Numerical state identifier column not entered but assumed.
1 1 0 0	Place one space between columns; Spaces between rows not required.
	—(Space required here)—
0 1 0 1	
0 0 0 0	Table for output T from Fig. 4.1(c) with inputs AB unfolded in *binary*.
0 0 0 0	Numerical state identifier column not entered but assumed.
0 0 0 0	Place one space between columns; Spaces between rows not required.

TABLE 4.1: (*continued*)
(c) The Following is the batch file Fig41c.bat for the SR model design of Fig. 4.1(c). adam1 /SR   Fig41c > Fig41c.out1 adam2 /SR   Fig41c > Fig41c.out2 adam3 /SR   Fig41c > Fig41c.out3 adam4 /SR   Fig41c > Fig41c.out4 adam5 /SR   Fig41c > Fig41c.out5 adam6 /SR   Fig41c > Fig41c.out6 adam7 /SR   Fig41c > Fig41c.out7

necessary to overcome the DOS barrier when designing large STT FSMs. For our purposes, we will design the STT FSM in Figure 4.1c to operate with C-elements requiring use of the SR batch file Fig41c.bat given Table 4.1.

After the input and batch files shown in Table 4.1 are constructed using a .txt editor, they must be saved to the C:\ADAM directory. Once they are there, the batch file can be activated (e.g., double click on Fig41c.bat), after which the seven output file contents given in Table 4.1c will be generated and will appear in the C:\ADAM directory. Any of these files can be viewed through the *edit* feature. The most useful of the output files are .out1 and .out7. The major parts of .out1 are given in Table 4.2 and are seen to be in agreement with Eqs. (4.1). The main parts of .out7 are given in Table 4.3. Here, it will be seen that the cubes for S_2, R_2, S_1, R_1, S_0, R_0 are not necessarily the same as those given in Eqs. (4.6) but they are, nevertheless, valid. If you consider that ADAM must minimize (with ESPRESSO) state assignment maps such as those in Figure 4.1d with several don't care states, it is expected that ADAM will generate one of a variety of valid NS logic solutions. In logic design, it is well understood that the larger the number of don't care states in a logic map to be minimized, the greater the number of possible valid design solutions. For further details on this and other relevant subjects, the reader should consult the *Readme.doc* provided with ADAM. It will become evident that ADAM is truly a versatile and powerful design tool.

A simulation of the ADAM SR state equations is identical with that in Figure 4.4 with the exception of a function hazard that occurs in the $y_1(H)$ state variable, and a static hazard that is produced in $T(H)$. Function hazards are not unusual and must be expected when using ADAM for design. This software will normally generate more p-terms (redundant PIs) than necessary in an attempt to maximize the number of shared PIs, some subject to function hazard production. Reading the logic for output T from Table 4.3a gives

$$T = \bar{y}_2\bar{y}_0 AB + \bar{y}_1 y_0 AB \qquad (4.7)$$

TABLE 4.2: ADAM results given in the file .out1 showing only the π- and τ-partitions, state code assignments, and the destination matrix.

The Following lists all the π-Partitions	The Following is the State Assignments Matrix as Derived from the τ-Partitions
0001 0	
01X0 1	000
0X10 2	001
X01X 3	011
X10X 3	110
0011 4	

The Following are all the τ-partitions as Derived from the π-Partitions	The Following is the Destination Matrix as Derived from the State Table:
0001 0	0000 0000 1XX1 11XX
0110 2 1	1111 0000 X1XX 0000
0100 3 1	0000 111X XX1X 0000
0010 3 2	0000 XXX1 0000 XX11
0011 4 3	

where it is clear that an internally initiated static-1 hazard exists during the $b \rightarrow a$ ($001 \rightarrow 000$) when $y_0 \rightarrow \bar{y}_0$ under branching conditions AB. Hazard cover for this S-hazard is $\bar{y}_2\bar{y}_1 AB$. Note that this hazard could be avoided by using the logic $T = \bar{y}_1 AB$ as given in Eqs. (4.5). See Section 3.3.2 for a review of static hazards in the output forming logic. Simulation of the p-term table in Table 4.3b for PLA design yields exactly the same logic waveforms. However, the static-1 hazard in the output $T(H)$ still remains as expected by considering the last two p terms in Table 4.3b and Eq. (4.7).

The reader should see from an inspection of the SR and p-term cubes in Table 4.3 together with the state diagram in Figure 4.1c that the minimum requirements for E-hazards #1 and #2 discussed in Section 4.3 are met for either implementation of this asynchronous FSM.

4.5   SUMMARY OF HAZARD EFFECTS AND THEIR ELIMINATION IN STT FSM DESIGNS

The following summary is presented to help the reader put into perspective the effects of hazard formation and elimination in the design of asynchronous FSMs.

TABLE 4.3: (a) ADAM results given in the file .out7 showing only the final SR state equations and the output equations, and (b) P-term table in Berkeley format for a PLA design.

(a) CUBES SHOWN HAVE THE FORM: $y_2 y_1 y_0 A B$		(b) P-TERM TABLE IN BERKELEY FORMAT FOR PLA DESIGN	
The Final Next State Equations in SR form are:		*Table*	*Comments*
$S_2 = {-}1{-}11$	The Final Output Equations are:	.i 6	(Number of inputs)
		.o 6	(Number of outputs)
$R_2 = {-}0{-}11, {----}0$	$R = {-}0110$	.ilb y2 y1 y0 A B SAN	(Input names)
		.ob Y2 Y1 Y0 R S T	(Output names)
$S_1 = {-}0{-}01$	$S = 1{--}01$	.p 15	(Number of P-terms)
		-11101 011000	
		--1011 011000	
$R_1 = 0100{-}, {--}0{-}0, {---}00$	$T = 0{-}011{-}0111$	-01101 001000	
		-0-011 011000	
$S_0 = {-}0{-}01, {---}00$		---001 001000	
		-11-01 001000	
		--10-1 001000	
$R_0 = {-}1{-}11, {-}0{-}11$		-01-01 001000	
		1--011 110000	
		-1-111 110000	
		11--11 110000	
		-0110- 000100	
		1--01- 000010	
		0-011- 000001	
		-0111- 000001	
		.e	(End)

[1] *Static hazards* must be eliminated from the NS forming logic in asynchronous FSMs designed by using the LPD model. This refers to static-1 hazards since we emphasize the use of SOP NS forming logic in this text. Such hazards are filtered out by using the nested cell model particularly when C-elements are used as the memory. No such filtering mechanism is possible for the output forming logic in asynchronous FSMs for which hazard cover must

be added if it is determined that a static hazard in a given output can affect the operation of the next stage to which it is an input.

[2] There is no hazard cover that can be used to protect an asynchronous FSMs from the formation of *function hazards*, which are common in the NS and output forming logic of these machines. This is true independent of the Huffman design model used, LPD or nested memory cell. Remember that function hazard formation is the proper response of the NS and output forming logic to inputs that change close to each other. Although the use of C-elements in the memory stage can help filter out some function hazards, the only reliable means of eliminating function hazards is to make certain that such input changes are separated by some satisfactory minimum period. Arbiters can be used to provide this minimum separation, as explained in Chapter 11. What the designer does not want are function hazard transients to reach near or slightly exceed the switching threshold of a given gate where the effect on FSM operation can become unpredictable.

[3] *E-hazards* (sequential hazards) are potential timing defects that, if present and active, are certain to cause the malfunction of asynchronous FSMs in which they are formed in FSM designs that operate in the fundamental mode. This includes FSMs that use C-elements as memory in designs we call quasi-Muller FSMs. These timing defects are produced by delays placed at specific places in an asynchronous FSM with delay magnitudes that exceed certain minimum values. The problem is that most of these delays are unintended and may occur as a result of foundry errors in chip production or that may occur simply due to transmission line delays or both. In modern high-speed CMOS logic designs, the minimum delays required to produce E-hazards are quite small, and may be of the order of a nanosecond. The only reliable means of eliminating the effects of a given E-hazard is to place a counteracting delay of some safe value on the feedback line of the second invariant state variable. This is explained in detail in Section 3.6.

[4] Chapters 8 and 9 in Part II of this text feature Muller-type systems that effectively deal with all hazard and race-related timing defects in the NS forming logic.

CHAPTER 5

Design of One-Hot Asynchronous FSMs

The one-hot design of asynchronous finite state machines (FSMs) requires a state code assignment consisting of a single "1" per state. One possible four-state state code assignment would be 0001, 0010, 0100, 1000, where a transition between any two states represents a Hamming distance of 2. Thus, n states each require n state variables. In a one-hot circuit, each state-to-state transition requires that the active "1" of the origin state remain active until the transition to the destination state is complete. As a consequence, this forces the FSM to transit through a state with two 1's, a state consisting of 1's from both the origin and destination states. This process continues as the asynchronous FSM undergoes the remainder of its allowable transitions. The reader will learn that this approach to asynchronous FSM design is remarkably simple requiring no next state (NS) or output K-map optimization or static hazard analysis.

5.1   INTRODUCTION TO THE ONE-HOT APPROACH

Each NS function in a one-hot design of an asynchronous FSM is represented generally in Eqs. (5.1), given as follows and consisting of two parts: the "into" terms and a single "out of" term.

$$\left\{ \begin{array}{l} Y_j = \underbrace{\sum_{k=0}^{m-1} y_k f_{j \leftarrow k}}_{\text{"Into" Terms}} + \underbrace{y_j \bar{F}_j}_{\substack{\text{"Out of"} \\ \text{Term}}} \\ Z_\ell = \sum_{k=0}^{m-1} y_j f_{j,\ell}(X) \end{array} \right\} \qquad (5.1)$$

Here, the "into" terms of Eqs. (5.1) represent the Boolean sum of all $y_k f_{j \leftarrow k}$ branching conditions *into* each jth state of a one-hot FSM. The "out of" term for each NS variable Y is responsible for forcing the FSM to transit via a state with two 1's during an origin-to-destination state transition. F_j in Eqs. (5.1) is the Boolean sum of all y variables in states to which the jth state transits, and $\bar{F}_j$ is the complement of that sum. The model for the "into" terms to the jth state of Eqs. (5.1) is shown

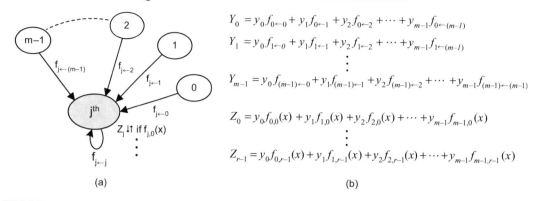

FIGURE 5.1: Model for the "into" terms of Eqs. (5.1). (a) State diagram segment showing "into" terms branching conditions and Mealy outputs for the jth reference state. (b) Generalized one-hot NS and output forming logic for the "into" terms given in Eqs. (5.1) for m states and r total outputs.

in Figure 5.1a, together with the output function for that state as generalized in Eqs. (5.1). The NS and output functions are given in Figure 5.1b, where only the "into" terms are represented for the NS functions.

The r output expressions in Z_ℓ of Eqs. (5.1), summed over m states, represent conditional (Mealy) outputs. Here, $f_{j,\ell}(X)$ represents the jth function of external inputs X for the ℓth output with $\ell = 0, 1, 2, \ldots (r-1)$. For unconditional (Moore) outputs, $f_{j,\ell}(X) = 1$.

5.2   CHARACTERISTICS OF THE ONE-HOT METHOD

Before presenting an example of the one-hot method, it is instructive to briefly discuss the salient features of the one-hot method.

- State code assignments are *not* needed or desired. Instead, use is made of state identifiers, either alphabetical (*a*, *b*, *c*, etc.), or numerical (0, 1, 2, etc.) used as (y_0, y_1, y_3,...) in the NS and output functions.
- The NS and output equations are read directly from a state diagram or state table—m states require m NS Y functions. This means that there will be m PS feedback paths required to implement a one-hot design. For very complex FSMs, the large number of feedback paths may be prohibitive.
- All *cycles* and *buffer states* must be eliminated prior to any one-hot design. Thus, every state must have a *holding condition* that will not allow further transitions until an input changes.
- Static hazards in the NS functions are always *internally initiated* but are covered by the holding conditions of the state. Static hazards in the output logic are not possible.

- Initialization is accomplished by the "1-hot + zero" method in which an all "0" state is initialized into the beginning one-hot state. Thus, a term $\bar{y}_0\bar{y}_1\bar{y}_2\ldots\bar{y}_{m-1}$ is added to one-hot NS variable of the initialized one-hot state. Combining the "out of" term and the "1-hot + zero" term reduces the latter by one $\bar{y}$ variable after applying the factoring and absorptive laws.

- The normal C-element implementation of a one-hot design is by the $Y \to S\bar{R}$ *one-hot conversion algorithm*:

$$\left\{\begin{matrix} S_j = \sum(\text{sum}) \text{ of all non-}y_j \text{ p terms in } Y_j \\ \bar{R}_j = \sum \text{ of all } y_j \text{ p-term coefficients in } Y_j \end{matrix}\right\} \qquad (5.2)$$

Here, the initializing *1-hot + zero* term is automatically included by the $Y \to S\bar{R}$ conversion algorithm. The conversion algorithm for the $\bar{R}_j$ p terms will also automatically include the necessary logic to cover the "out of" term requirements. Thus, if a set–reset (SR) design is required, it is strongly recommended that the lumped path delay (LPD) logic functions be produced first.

- Essential hazards (E-hazards) are always formed via ANDing race gates (RGs) and are highly predictable from the state diagram or state table.

5.3   DESIGN EXAMPLE USING C-ELEMENTS

To illustrate the one-hot FSM design method, we will take a simple example with the understanding that the relatively simple procedure involved is easily applicable to more complex FSMs. Shown in Figure 5.2a is a four-state FSM for which there are no cycles or buffer states, a requirement of the one-hot design. There are two input variables, A and B, and three outputs, W, X, and Z. Here, use is made of alphabetic state identifiers in place of the y_i notation in the NS functions. In doing this, function representation is significantly simplified. Figure 5.2b shows the NS functions written directly from the state diagram in Figure 5.2a. The "into" terms are easily understood from an inspection of Figure 5.1 where y_k are replaced by the state identifiers a, b, c, and d for simplicity. These state identifiers could represent 0001, 0010, 0100, and 1000, respectively—the choice of "1" orientations being optional. The "out of" term in each NS function is designed to force the FSM to retain the logic 1 of the origin state until the transition to the destination state is complete. Thus, the FSM transitions through a state having two 1's in passage from the origin to the destination state. For example, the state a "out of" term takes the form $\bar{a}\bar{b}\bar{c}$ because state a transits to both state b and state c. According to Eqs. (5.1), the *1-hot + zero* term for this FSM must be $\bar{a}\bar{b}\bar{c}\bar{d}$. When combined with the "out of" term $\bar{a}\bar{b}\bar{c}$, the two terms become $a\bar{b}\bar{c} + \bar{a}\bar{b}\bar{c}\bar{d} = \bar{b}\bar{c}(a + \bar{a}\bar{d}) = a\bar{b}\bar{c} + \bar{b}\bar{c}\bar{d}$ by using the factoring and absorptive laws. (See Appendix A.2)

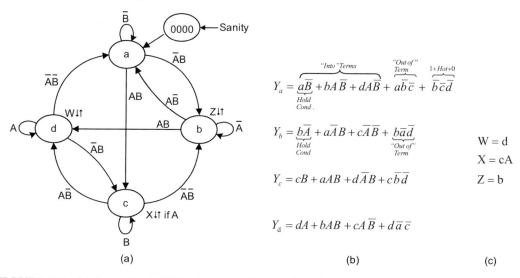

$$Y_a = \underbrace{a\bar{B}}_{\substack{Hold \\ Cond.}} + bA\bar{B} + dA\bar{B} + \overbrace{ab\bar{c}}^{\substack{\text{"}Out\,of\text{"} \\ Term}} + \overbrace{\bar{b}\bar{c}\bar{d}}^{1+Hot+0}$$

"Into" Terms

$$Y_b = \underbrace{b\bar{A}}_{\substack{Hold \\ Cond}} + a\bar{A}B + c\bar{A}\bar{B} + \underbrace{\bar{b}\bar{a}\bar{d}}_{\substack{\text{"}Out\,of\text{"} \\ Term}}$$

$$Y_c = cB + aAB + d\bar{A}B + c\bar{b}\bar{d}$$

$$Y_d = dA + bAB + cA\bar{B} + d\bar{a}\bar{c}$$

W = d

X = cA

Z = b

FIGURE 5.2: (a) A one-hot FSM having no cycle paths but showing the 0000 state as required by the one-hot-plus-zero initialization method. (b) The one-hot LPD logic for the FSM in (a). (c) Output logic.

By using the $Y \rightarrow S\bar{R}$ *one-hot conversion algorithm* given by Eqs. (5.2), the LPD NS equations in Figure 5.2b are converted to the SR logic in Eqs. (5.3) suitable for use with normal C-elements of the type shown in Figure 1.12 and illustrated in Figure 2.2b. Normal C-elements are used because the $Y \rightarrow S\bar{R}$ conversion algorithm automatically generates the necessary $\bar{R}_j$ requirement for each C-element input. Presented in Figure 5.3 is the NAND/C-element implementation of Eqs. (5.3). The *1-hot + zero* term is represented by the shaded gate. Note that the NS logic gate tally for the LPD and SR designs are the same (21), but without taking account of C-element use for the SR design.

$$\overbrace{1- Hot + 0}$$

$$S_a = bA\bar{B} + d\bar{A}\bar{B} + \bar{b}\bar{c}\bar{d} \qquad \bar{R}_a = \bar{B} + \bar{b}\bar{c}$$

$$S_b = a\bar{A}B + c\bar{A}\bar{B} \qquad \bar{R}_b = \bar{A} + \bar{a}\bar{d}$$

$$S_c = aAB + d\bar{A}B \qquad \bar{R}_c = B + \bar{b}\bar{d} \qquad (5.3)$$

$$S_d = bAB + cA\bar{B} \qquad \bar{R}_d = A + \bar{a}\bar{c}$$

The simulation of the C-element logic circuit in Figure 5.3 is shown in Figure 5.4a. A careful inspection of the waveforms indicates that the sequential behavior required by the state diagram

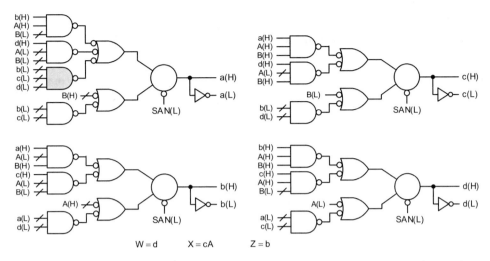

$$W = d \qquad X = cA \qquad Z = b$$

FIGURE 5.3: C-element implementation of Eqs. (5.2) showing the 1-Hot+0 state (shaded), the necessary SANITY(L) input, and the three output functions given in Figure 5.2.

in Figure 5.2a is followed. However, it is also obvious that each origin-to-destination state transition takes place via a state containing the two 1's of these states. As discussed earlier, this is a direct result of the "out of" term found in each NS function. This is emphasized in Figure 5.4b, which is a blowup of the shaded region in Figure 5.4a. As a result, overlapping outputs are possible (*W*

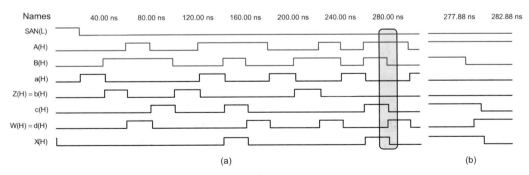

FIGURE 5.4: (a) Simulation of the one-hot FSM in Figure 5.2a implemented with the C-element circuit in Figure 5.3 and in agreement with Eqs. (5.3). (b) Blowup of the shaded area in (a) showing the *c*-to-*d* transition via state 0011 where output $X = cA$ is maintained active with *c*.

and X in this case) so care must be exercised when using such output signals as inputs to any next stage. Note that there are no static hazards present in this design. The static hazards that exist in the LPD Y_j functions given in Figure 5.2 are formed between an "into" term and the "out of" term, but are each covered by the holding condition for the function. Thus, static hazards are absent in any one-hot design. This is not true for essential hazards as is discussed in the next section.

5.4 ESSENTIAL HAZARDS IN ONE-HOT ASYNCHRONOUS FSMs

As with any asynchronous FSM of three or more states, E-hazards will almost certainly be present. In one-hot designs, E-hazards are particularly easy to analyze and are highly predictable. In fact, the RGs for E-hazards in one-hot asynchronous FSMs are *always* ANDing RGs. Following Section 3.6 and using the state diagram in Figure 5.2a, we will now analyze the two E-hazards that exist in this FSM, one $a \rightarrow c \rightarrow d$ and the other $b \rightarrow d \rightarrow c$. Both of these E-hazards satisfy the minimum requirements for E-hazard formation given in Figure 3.9. To carry out this analysis, we will need to take into account the two cycle-paths involved for each E-hazard, each cycle path being from one one-hot state to another. We will use Eqs. (5.3) for the analysis.

E-Hazard #1. The required path for this E-hazard is $a \rightarrow c \rightarrow d$ under input conditions $A\bar{B} \rightarrow AB$. The appropriate cycle paths for this E-hazard are given below where the first and second invariants are identified as y_3 under AB, and y_2 under $A\bar{B}$, respectively. An inadvertent delay Δt_{E1} of sufficient magnitude must exist on the initiator line B to y_3. However, it is not necessary to use the y notation. Consistent with the equation in Figure 5.2b, we assign $y_3 = d$ and $y_2 = c$, etc. Thus, $a \rightarrow (ac) \rightarrow c$ means $0001 \rightarrow 0101 \rightarrow 0100$ with d (first bit, y_3) as the first invariant. Then, c is the second invariant as shown below. So we deduce that the ANDing RG for this E-hazard must be $cA\bar{B}$ in S_d.

$$\overbrace{0001}^{a} \rightarrow (0101) \rightarrow \overbrace{0100}^{c} \rightarrow (1100) \rightarrow \overbrace{1000}^{d}$$
$$\underbrace{\qquad}_{AB} \qquad \underbrace{\qquad}_{A\bar{B}}$$

The indirect path (IP) must not be inconsistent with a, A (a constant), and must contain B or $\bar{B}$ in c. Therefore, the IP must be via aAB in c (i.e., in S_c). With this information, it becomes clear that the minimum requirements for E-hazard #1 to form must be $\Delta t_{E1} + \tau_{Int} > (\tau_{aAB} + \tau_{C\text{-element}})$. Should this E-hazard form, the FSM would transit $a \rightarrow (0101) \rightarrow (1100)$ and reside stably in the 1100 cycle state, never completing the transition to state d, a serious malfunction of the FSM.

Once the minimum path delay requirements for E-hazard formation are known, a reasonable choice can be made for the delay $\Delta t_{Correct}$ that could be placed on the c line to the RG. Once

$\Delta t_{Correct}$ is properly placed, the probably of formation of E-hazard #1 would be greatly diminished. Naturally, $\Delta t_{Correct}$ need not be used if the designer feels confident that this E-hazard cannot form. Of course, such guesswork could become a serious mistake.

E-Hazard #2. The required path for this E-hazard is $b \rightarrow d \rightarrow c$ under input conditions $\overline{A}B \rightarrow AB$ with cycle paths given by

$$\overbrace{0010}^{b} \rightarrow \underbrace{(1010) \rightarrow \overbrace{1000}^{d}}_{AB} \rightarrow \underbrace{(1100) \rightarrow \overbrace{0100}^{c}}_{\overline{A}B}$$

With this information, we conclude that the first and second invariants are c (under AB) and d (under $\overline{A}B$), respectively. Thus, the ANDing RG must be $d\overline{A}B$ in c (i.e., in S_c). To activate this E-hazard, an inadvertent delay Δt_{E2} of sufficient magnitude must exist on the initiator line A to the first invariant c. The IP must not be inconsistent with b, constant input B and must contain A or $\overline{A}$ in d. Thus, the IP is via bAB in d (i.e., in S_d). From this information, we deduce that the minimum delay requirements for E-hazard #2 formation is $(\Delta t_{E2} + \tau_{Inv}) > (\tau_{bAB} + \tau_{C\text{-element}})$. To be effective, the counteracting delay $\Delta t_{Correct}$ must be placed on the c line to the RG.

CHAPTER 6

Design of Pulse Mode FSMs

Up to this point, we have dealt exclusively with asynchronous design methods that permit input signals to overlap but with the potential to have certain timing defects that could cause the asynchronous finite state machine (FSM) to malfunction. Asynchronous state machines that are designed to operate with nonoverlapping pulsed inputs and that operate with "data triggered" memory elements are called *pulse mode* sequential machines. The pulse mode approach offers a simple means of designing asynchronous FSMs but at the price of greatly restricting input signal conditions. The pulse mode approach eliminates the timing defects owned exclusively by fundamental mode FSMs, and eliminates the problems associated with clock skew and the need to synchronize inputs to the clock. Thus, at first glance, it would seem that the pulse mode approach to FSM design has all of the advantages of synchronous FSMs while being free of the timing defects common to fundamental mode (Huffman) machines or Muller machines. Although this is true, the apparent advantages of the pulse mode methodology are offset by the severe limitations placed on the input data signals.

6.1   MODELS AND CHARACTERISTICS OF THE PULSE MODE

Shown in Figure 6.1 is the general Mealy model for the an asynchronous FSM designed by using the pulse mode approach. Note that data-triggered toggle modules are used as memory elements. Thus, one major difference between the pulse mode model and that of the lumped path delay (LPD) and nested cell models in Figures 1.1 and 1.3 is the memory stage. Recall that in the LPD model of Figure 1.1 the memory stage was made up of fictitious LPD memory elements from which the stability criteria of Eqs. (1.1) and (1.2) and excitation table in Figure 1.2 are derived. The nested element model of Figure 1.3 requires the use of either C-elements or basic cells for which the $Y \rightarrow SR$ conversion is used. Toggle modules, on the other hand, are similar to T flip-flops but with a single enabling input called T into which data are entered as discrete positive pulses. It is these discrete pulse inputs to T that triggers the toggle modules, thereby controlling the data flow through the sequential machine.

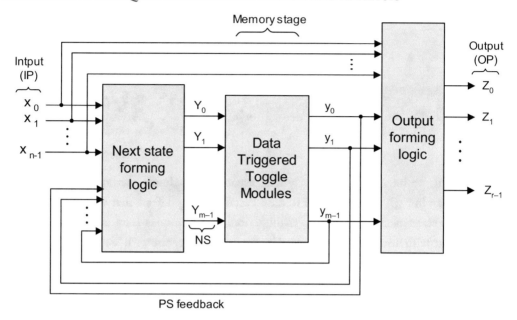

FIGURE 6.1: The generalized model for an asynchronous Mealy FSM operated in the pulse mode with data-triggered toggle modules as the memory elements.

6.1.1   Requirements and Characteristics of the Pulse Mode Approach

There are a number of unique requirements and characteristics that set pulse mode FSMs apart from the fundamental mode approaches to asynchronous FSM design discussed previously in this text. These are listed below but not in any specific order of importance.

1. Branching conditions in a pulse mode state diagram are always *single variables* or *ORed single variables* (e.g., $X + Y$) that are always uncomplemented as required for positive pulses. Unconditional branching in a pulse mode state diagram is strictly forbidden.
2. Any state coding will suffice. However, owing to the toggle character of the toggle module memory elements, a binary sequence is preferred wherever possible so as to minimize the next state (NS) logic. Recall the toggle character of the binary code beginning with the least significant bit and proceeding all the way to the MSB.
3. Data must be presented to a pulse mode circuit as discrete nonoverlapping positive pulses that are at least minimally separated as shown in Figure 6.2. Although there is in no upper bound place on their active duration, there is a required lower bound.

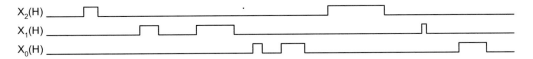

FIGURE 6.2: Examples of nonoverlapping data input signals that are at least minimally separated positive pulses having active durations with no upper bound.

4. The NS logic is obtained by combining the excitation table for the toggle module (T flip-flop) with the pulse mode state diagram by using the mapping algorithm in Section 1.6.

5. Because states in a pulse mode design cannot toggle to themselves, only outgoing single variable or ORed single variable branching conditions are required in mapping the NS logic. Thus, holding condition should not be given in the state diagram or state table.

6. From the characteristic described above, it is obvious that the *sum rule* is never obeyed but the mutually exclusive requirement is uniquely satisfied by the nonoverlapping input requirement.

7. Falling edge triggered (FET) toggle modules must be used with the discrete positive pulses as in Figure 6.2.

8. When it is appropriate to do so, the outputs should be made conditional on the exciting branching variable (a Mealy output) when using FET toggle modules. Doing this results in two important advantages that derive from falling edge triggering:
   (a) Output race glitches (ORGs) are not possible.
   (b) Static hazards in the output forming logic are not possible.
   These advantages are not guaranteed if unconditional (Moore) outputs are used.

9. Initialization methods are exactly the same as used for LPD model designs or for C-element designs discussed in Section 2.2. When using a C-element design, initialization must be into the all "0" state.

10. Proper pulse mode designs cannot have endless cycles, critical races, or E-hazards. Static-1 hazards in the NS logic are not possible when FET toggle modules are used.

11. Debouncing of inputs from mechanical switches is imperative because pulse-mode circuits are highly sensitive to transient signals of sufficient duration and strength.

12. Synchronization of inputs to a pulse mode FSM is not necessary because the requirement of nonoverlapping data inputs, at least minimally separated, is a form of synchronization.

13. Pulse mode FSMs that are properly designed and operated cannot go metastable and, therefore, have an infinite *mean time between failures*.

The 13 characteristics of the pulse mode approach to asynchronous FSM design just given should seem impressive compared to the LPD and nested element approaches that operate in the fundamental mode. It would appear that pulse mode FSMs have all the benefits of asynchronous fundamental mode FSMs but without their inherent problems. This is true! However, the price that must be paid for this "perfection" is the sever restrictions placed on the input signals and the memory elements use to process them—restrictions that require discrete nonoverlapping pulses at least minimally separated to be the inputs to FET toggle modules.

6.2   TOGGLE MODULES AS THE MEMORY ELEMENTS

Toggle modules function such as T flip-flops or D flip-flops have been connected in such a manner as to make them operate as toggle modules. Actually, D flip-flops can be converted to T flip-flops or to toggle modules. See Tinder's text in Endnotes for a detailed discussion of these subjects. However, it is not necessary to use flip-flop conversion to obtain toggle modules that can be designed directly and more simply from first principles. Shown in Figure 6.3 is the design of a FET toggle module by using either the LPD model or nested element model. Figure 6.3a shows the four-state state diagram representing a toggle module that can be operated as either an FET or RET (rising edge triggered) memory element—FET is preferred. Figure 6.3b illustrates the EV K-maps and logic for the LPD model design, and the $Y \rightarrow SR$ converted K-maps and logic for the nested element

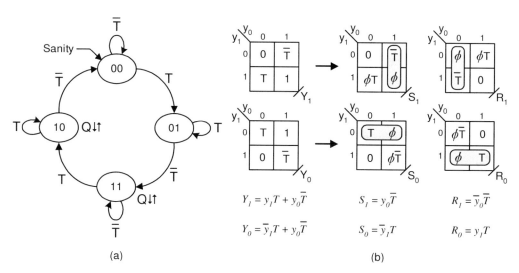

$$Y_1 = y_1 T + y_0 \overline{T}$$ $$S_1 = y_0 \overline{T}$$ $$R_1 = \overline{y}_0 \overline{T}$$

$$Y_0 = \overline{y}_1 T + y_0 \overline{T}$$ $$S_0 = \overline{y}_1 T$$ $$R_0 = y_1 T$$

(a)   (b)

FIGURE 6.3: FET toggle module design for use with the pulse mode circuits. (a) State diagram for the FET toggle module. (b) LPD logic and K-map conversion and logic for C-element or SR basic cell design of toggle modules.

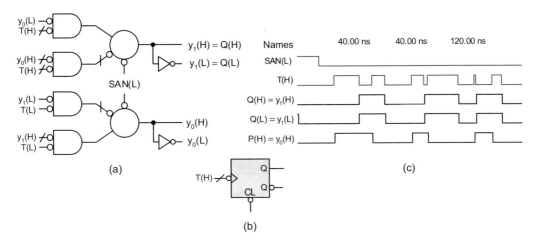

FIGURE 6.4: (a) Implementation of the toggle module with NOR gates and complementary C-elements. (b) Circuit symbol for a falling edge triggered (FET) toggle module with a T(H) data input. (c) Simulation of the circuit in (a) showing both falling edge triggerig (Q) and rising edge triggering (P).

design by using either C-elements or SR basic cells. See Section 2.1 for a review of the extended mapping algorithm required for the $Y \rightarrow SR$ conversion.

By using the logic given in Figure 6.3b, we have opted to design this toggle module by using NOR gates and complementary C-elements of the type shown in Fig 1.13. The resulting logic circuit is presented in Figure 6.4a, which is initialized into the 00 state as indicated by the SAN(L) input to the C-elements. The circuit symbol for the FET toggle module is given in Figure 6.4b. The simulation of the logic circuit in (a) is provided in Figure 6.4c with Q and P outputs for falling and rising edge triggering, respectively. For use in a pulse mode design with nonoverlapping positive pulses, it is the $Q(H) = y_1(H)$ or $Q(L) = y_1(L)$ outputs that must be used consistent with the logic circuit symbol in (b). An inspection of the waveforms in (c) indicates that the $Q(H)$ and $Q(L)$ output changes are indeed triggered by the falling edges of the $T(H)$ waveform. The waveform for $P(H) = y_0(H)$ changes with the rising edge of the $T(H)$ waveform and therefore defines an RET toggle module. Again, it is important for the reader to understand that the use of the Q outputs from the toggle module are essential to the proper operation of a pulse mode circuit design. They eliminate ORGs and static hazards in the output forming logic that might otherwise form with rising edge triggering.

6.3   A DESIGN EXAMPLE

The design of pulse mode FSMs by using toggle modules requires that the mapping algorithm in Section 1.6 be used together with the excitation table for the pulse mode model (the T column)

	PS variable change		LPD Model	Pulse Mode Model
	y_t	$\rightarrow$ y_{t+1}	Y_t	T
Reset Hold	0	$\rightarrow$ 0	0	0
Set	0	$\rightarrow$ 1	1	1
Reset	1	$\rightarrow$ 0	0	1
Set Hold	1	$\rightarrow$ 1	1	0

Toggle (for Set and Reset rows)

FIGURE 6.5: Excitation tables for the LPD model and pulse mode designs of asynchronous FSMs.

given in Figure 6.5. Thus, the pulse mode model requires the toggle conditions $0 \rightarrow 1$ and $1 \rightarrow 0$ be used in a manner similar to the two set conditions ($0 \rightarrow 1$, $1 \rightarrow 1$) used for the LPD model as discussed in Section 1.7. Keeping in mind this information and the characteristics of the pulse mode given in Section 6.1.1, we can now proceed with a simple pulse mode design.

The state diagram for a 2-bit digital combinational lock is shown in Figure 6.6(a). Note that there are no holding conditions in the state diagram, and that all branching conditions are single uncomplemented input variables or are ORed combination of the single uncomplemented variables. Clearly, the sum rule cannot hold for a pulse mode state diagram. To minimize the NS forming logic, the state code assignments are in binary sequence wherever possible. The output OPNVLT (for open vault) is conditional on the exiting input X. Thus, all requirements outlined in Section 6.1.1 are met.

The EV K-maps for the NS and output forming logic of the 2-bit digital combinational lock are given in Figure 6.6b. The NS K-maps are plotted by using the T excitation table, given in Figure 6.5, together with the mapping algorithm in Section 1.6. Thus, the "T" K-maps are plotted by toggle branching from each state. For example, in state 011, state variable y_1 toggles to state 000 under X but also toggles to state 100 under Y. This requires the entry to be $X + Y$ for state 011 in the T_1 K-map. There are three don't care (unused) states each indicated by a ϕ symbol. The resulting logic for the NS and output functions are given by Eqs. (6.1), where optimum use of the three don't care states is indicated by the shaded loops. Note the shared prime implicants (PIs) $y_1 y_0 Y$ and $y_0 X$ between the NS functions T_2 and T_1 and between T_1 and T_0, respectively, and the shared PI $y_2 X$ between T_2 and the output function OPNVLT. This gives a gate/input tally of 9/21 for the NS and output functions.

$$T_2 = y_2 X + y_2 Y + y_1 y_0 Y \quad T_1 = y_1 X + y_0 X + y_1 y_0 Y \quad T_0 = \bar{y}_2 Y + y_0 X$$

$$\text{OPNVLT} = y_2 X \qquad (6.1)$$

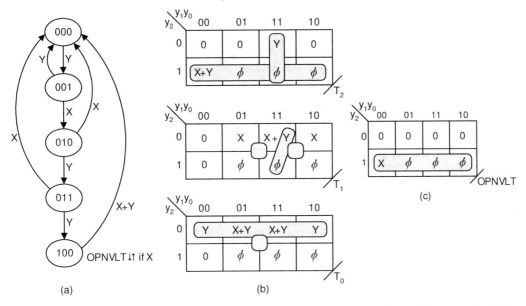

FIGURE 6.6: Design of a 2-bit digital combinational lock by using the pulse mode approach. (a) State diagram with a conditional (Mealy) output OPNVLT (for opten vault). (b) EV K-maps for use of toggle modules as the memory elements. (c) EV K-map for the conditional output, OPNVLT if X.

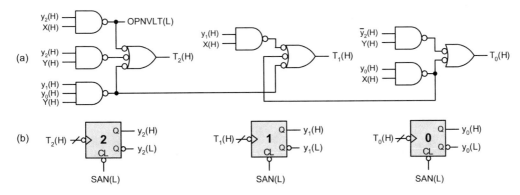

FIGURE 6.7: Design of the 2-bit digital combinational lock given in Figure 6.6. (a) Implementation of the NS and output forming logic given by Eqs. (6.1). (b) Toggle modules as the memory stage with initialization into the 000 state.

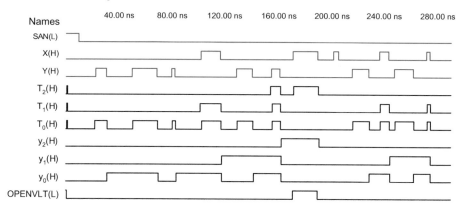

FIGURE 6.8: Simulation of the pulse mode FSM in Figure 6.7 designed to operate as a 2-bit digital combinational lock using toggle modules as the memory elements.

The NS and output functions given by Eqs. (6.1) are implemented with NAND logic in Figure 6.7a, where all three shared PIs are wired in. Note that the output is issued active low, OPNVLT(L) for convenience. The memory stage is made up of three FET toggle modules, as indicated in Figure 6.7b. The active low indicator bubble on the left side of each toggle module is indicative of the falling edge triggering of the memory module.

The simulation of the logic circuit in Figure 6.7 is given in Figure 6.8, where an initial Sanity(L) input 1(L) initializes the FSM into the 000 state. Then, after a 1(L) → 0(L) change in the Sanity(L) signal, the FSM operates properly. Here, the three $T_i(H)$ outputs are included for completeness, but it is clear that the PS state functions $y_i(H)$ change only on the falling edge of the data inputs X and Y. Previously, it was indicated that there is no upper bound on the active duration of the data pulses. It was also indicated that a lower bound does exist. If a data pulse is not developed sufficiently, it will not be read by the toggle module. Worse yet, if the data pulse is two weak to completely cross the switching threshold, it cannot be read predictably and a metastable condition may ensue.

6.4   OTHER MEMORY ELEMENTS SUITABLE FOR PULSE MODE DESIGN

Toggle modules of the type designed in Figures 6.3 and 6.4 are not the only type that can be used for pulse mode FSM design. We chose to use the C-element design in Figure 6.4 because C-elements are less susceptible to metastable conditions compared to, say, basic cell or LPD approaches. Still, there are other memory devices incorporating C-elements that can be converted into toggle mod-

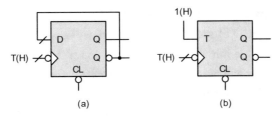

FIGURE 6.9: Alternative memory elements for use in asynchronous pulse mode FSM designs. (a) An FET D-flip-flop (FF) converted to an FET toggle module. (b) An FET T-FF converted to an FET toggle module.

ules. Shown in Figure 6.9 are two examples of alternative memory elements suitable for pulse mode FSM design. Figure. 6.9a features an FET D-FF that is converted to an FET toggle module by connecting its active low output to the D input. A T-FF can be converted to an FET toggle module by connecting its T input to logic $1(H) = HV$ as in Figure 6.9b. See Tinder's text in Endnotes for further details.

Generally speaking, the toggle module in Figure 6.4 is the most desirable simply because it is fast, requires the least amount of hardware for implementation, and is the most reliable option. Toggle modules converted from D-FF and T-FF require more logic hardware but are also reliable if implemented with C-elements. In any case, runt pulses should be avoided because they may or may not be of sufficient strength to cross the switching threshold.

6.5   DEBOUNCING CIRCUITS

Pulse mode FSMs are particularly susceptible to input transients that can be caused by mechanical switch bounce action. Such transients can cause a pulse mode FSM to transit through an unpredictable

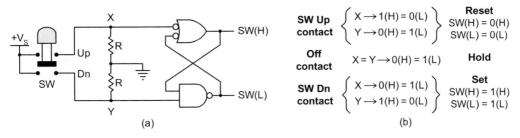

FIGURE 6.10: Debouncing the single-pole/double-throw (SPDT) switch by using a set-dominant basic cell in Figure 1.8. (a) Logic circuit. (b) Logic values for Up-, Dn- and Off-contact positions of switch, SW.

number of states or may even cause metastability in the FSM. To counteract bounce transients, a single-pole/double-throw (SPDT) switch can be effectively used, as indicated in Figure 6.10. If a spring-loaded switch (SW) is initially in the UP position, the SW outputs are in a Reset state. When the SW is depressed, it will pass through a Hold state before reaching the Dn contact. Upon reaching the Dn position, the first transient to cross the switching threshold will force the SW outputs into a Set state thereby avoiding any further transient action. The reverse process follows the release of the SW. Accordingly, the SPDT switch can never undergo an X,Y $1(L) \rightarrow 0(L)$ oscillation as in Figure 1.10.

CHAPTER 7

Analysis of Asynchronous FSMs

Analysis of asynchronous finite state machines (FSMs), as interpreted in this chapter, is a "reverse engineering" process. By this, we mean that the analysis process generally begins with a circuit and ends with a state diagram or state table. Of course, we have already analyzed circuits for various timing defects, and that is, or should be, part of any analysis performed on an asynchronous FSM. But remember that the state diagram of an asynchronous FSM serves as a most powerful tool in any analysis process. Not only can we analyze a state diagram for various timing defects, but many design irregularities become obvious sometime with only a cursory observation of the state diagram. We will hold to these notions throughout this chapter, with the result of achieving a practical and reliable means of analyzing any given asynchronous FSM.

7.1 PROCEDURE FOR ANALYZING ANY ASYNCHRONOUS FSM

The following three-part procedure should be followed in analyzing any asynchronous FSM with the assumption that a suitable starting point is available—namely, a logic circuit or some closely related representation that can be converted to a logic circuit.

Part 1: Begin with a logic circuit in mixed-logic form. If the logic circuit is presented in positive logic form, then convert it to mixed-logic form. This means that voltage levels and positive logic gate symbology must be converted to the mixed-logic form as used in this text. Attempting to do otherwise increases greatly the probability for error.

Part 2: Map the NS and output forming logic into lumped path delay (LPD) entered variable (EV) K-maps such that each cell has as its coordinates the present state (PS) variables, y_i, and as its contents a NS subfunction.

a) If the next state (NS) logic for the asynchronous FSM has been designed by using Set–Reset (SR) memory elements, read and map the SR logic into EV K-maps and then use the reverse conversion $SR \rightarrow Y$ algorithm to produce LPD EV K-maps. Note that don't cares will not be present because they cannot be represented in a circuit.

b) If the NS logic for the asynchronous FSM has been designed by using toggle modules as memory elements, read and map the T logic into EV K-maps and then use $T \rightarrow Y$ K-map conversion to obtain the LPD EV K-maps. The algorithm for $T \rightarrow Y$ K-map conversion is as follows: For all that *IS NOT* y_i in the T_i K-map, transfer it directly to the same domain cells in the Y_i K-map. For all that *IS* y_i in the T_i K-map, transfer each cell entry complemented to the same domain cells in the Y_i K-map. Again, no don't care states are possible in the analysis.

Part 3: Once the logic circuit has been mapped into LPD EV K-maps, construct the PS/NS table. Do this for each state with inputs represented in *canonical-literal form* as a binary sequence covering all possible input conditions for that state. As an example, if a given cell depends on two inputs, X and Y, then four entries are necessary $(\overline{X}\overline{Y}, \overline{X}Y, X\overline{Y}, XY)$. This is necessary so as to avoid possible branching omissions. Each entry in the NS column is found by comparing the NS cell entries in the K-maps with the corresponding input conditions in the inputs column. When completed, the PS/NS table is precisely the tabular form of the state diagram. The state diagram obtained from the PS/NS table will have all states represented including possible don't care states, buffer states, and hang states.

7.2 EXAMPLE OF AN LPD MODEL FSM ANALYSIS

The LPD logic circuit to be analyzed is given in Figure 7.1 and is presented in mixed-logic form. It is known that E is an enable input, C is a clock input, M is a mode input, and P is pulse output. Reading the circuit and simplifying there results the NS and output forming logic given by Eqs. (7.1)

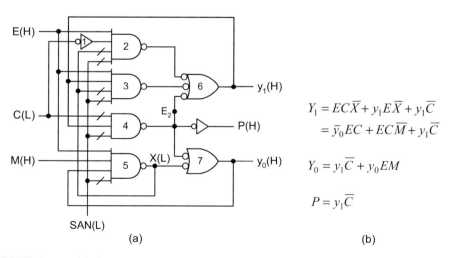

$$Y_1 = EC\overline{X} + y_1 E\overline{X} + y_1\overline{C}$$
$$= \overline{y}_0 EC + EC\overline{M} + y_1\overline{C}$$

$$Y_0 = y_1\overline{C} + y_0 EM$$

$$P = y_1\overline{C}$$

FIGURE 7.1: (a) A three-input, one-output LPD Logic circuit to be analyzed. (b) Eqs. (7.1) gives the NS and output forming logic as read and simplified from the circuit in (a).

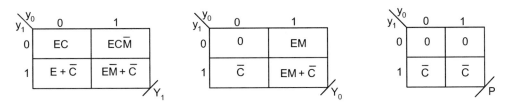

FIGURE 7.2: Next state and output EV K-maps as plotted from Eqs. (7.1) in Figure 7.1.

in Figure 7.1, where the intermediate function $X = y_0 EM$ giving $\bar{X} = (\bar{y}_0 + \bar{E} + \bar{M})$. At first glance, it would appear that there are two externally initiated static-1 hazards in function Y_1 when $C \to \bar{C}$ in state 10. However, the inclusion of $\bar{X} = (\bar{y}_0 + \bar{E} + \bar{M})$ into the first equation for Y_1 shows that these two hazards are covered by the terms $y_1 \bar{y}_0 E$ and $y_1 E\bar{M}$ inherent in Y_1. Thus, no static hazards exist in Eqs. (7.1). See Section 3.3 for a review of static hazard detection and elimination in the NS logic.

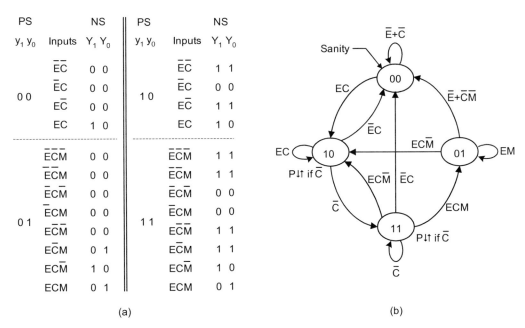

FIGURE 7.3: (a) PS/NS table obtained from the Y K-maps in Figure 7.2 and (b) the state diagram for the FSM represented by the logic circuit in Figure 7.1a as derived from the PS/NS table in part (a) and the output K-map in Figure 7.2.

The EV K-maps for the NS and output forming logic in Eqs. (7.1) are given in Figure 7.2. From these K-maps, the PS/NS table is constructed and presented in Figure 7.3a. Note that the inputs are listed in canonical-literal form in binary sequence covering all possible input combinations for each state. In state 00, for example, an inspection of the EV K-maps indicates that cell 00 (for state 00) depends only on inputs E and C. Therefore, the "Inputs" column in the PS/NS table corresponding to state 00 has only the four canonical-literal entries given as a binary sequence. In contrast, states (and K-map cells) 01 and 11 depend on inputs E, C, and M so there will be eight canonical-literal entries in the PS/NS table for these two states. It is true that shortcuts can be taken in constructing the PS/NS table. However, this is not recommended because the probability of branching errors is likely to increase.

The PS/NS table in Figure 7.3a is precisely the tabular form of the state diagram shown in Figure 7.3b. For example, in PS 00, the 00 → 00 holding condition is $\bar{E}\bar{C} + \bar{E}C + E\bar{C} = \bar{E} + \bar{C}$, and the 00 → 10 branching condition is EC. In state 01 the 01 → 10 branching condition is read simply as $EC\bar{M}$; the 01 → 01 (holding condition) is $E\bar{C}M + ECM = EM$. The 01 → 00 branching condition for the five entries is most easily found by constructing a conventional third-order 1's and 0's K-map with ECM as its coordinates. Looping out the five entries readily yields $\bar{E} + \bar{C}\bar{M}$ as the conditions for this branching path. The output function for this FSM is read directly from the EV K-map in Figure 7.2 and requires that a conditional output P if $\bar{C}$ be issued in states 10 and 11.

Now that we have obtained the state diagram for this FSM, it would be instructive to know exactly its function. A detailed inspection only of the state diagram in Figure 7.3b will most likely yield incomplete or incorrect information. However, combined with a simulation, the state diagram can contribute significantly to an understanding of its purpose. Shown in Figure 7.4 is the simula-

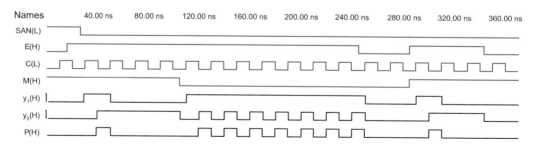

FIGURE 7.4: Simulation of the circuit in Figure 7.1 showing this FSM to be a pulse generation module whereby a single pulse or string of pulses are issued antiphase with clock (C) as controlled by the mode (M) and enable (E) inputs.

tion of this FSM. It appears that this FSM is a *pulse generation module* (PGM) that can issue a single pulse or a string of pulses, all antiphase to the clock input and all controlled by the enable and mode inputs E and M, respectively. If the PGM enters state 11 and then exits to state 01 on ECM or to state 00 on $\bar{E}C$, only a single pulse will be issued. On the other hand, if the PGM cycles with C between states 11 and 10 under $E\bar{M}$, then a string of pulses will occur, one pulse for each cycle but antiphase to the clock (C) input. If, at any time, the enable input E goes inactive $1(H) \rightarrow 0(H)$, then no pulses can be issued.

7.2.1   E-hazard and D-trio Analyses of the PGM

A close inspection of the state diagram in Figure 7.3b shows that there are two E-hazards and two d-trios. We will briefly consider each of these in turn and judge whether they pose a potential threat to the proper operation of the PGM. To simplify the analysis process, we will make use of the gate numbers indicated in Figure 7.1a. Also, before reading the following, the reader is advised to review Section 3.6.

1. The E_1-hazard path is $00 \rightarrow 10 \rightarrow 11 \rightarrow 01$ under input conditions $E\bar{C}M \rightarrow ECM$ due to an unintended delay of Δt_{E1} on the initiator C line to the ANDing race gate (RG) $y_1\bar{C}$ (gate 4) in y_0. The indirect path (IP) must not be inconsistent with $\bar{y}_1, \bar{y}_0, E, M$ and must contain the initiator as either C or $\bar{C}$ in y_1. Therefore, the IP must be via $EC\bar{X}$ (gate 2). We conclude that $\Delta t_{E1} > (\tau_1 + \tau_2 + \tau_6)$ as the delay requirement for E-hazard formation. Here, τ_1 is the propagation delay through an inverter commonly taken to be about $\frac{1}{3}\tau_p$, where we use τ_p to represent an average gate delay. Note that this E_1-hazard is the fist we have encountered that involved three transitions and four states.

2. The E_2-hazard path is $10 \rightarrow 11 \rightarrow 01$ under input conditions $ECM \rightarrow E\bar{C}M$ due to an unintended delay of Δt_{E2} at E_2 (see Figure 7.1) to ORing RG (gate 6) in y_1 via the initiator C path to $y_1\bar{C}$ (gate 4). The IP must not be inconsistent with $y_1, \bar{y}_0, E, M$ and must contain C or $\bar{C}$ in y_0. Therefore, $\Delta t_{E2} > (\tau_7 + \tau_5 + \tau_3)$ as the requirement to activate the E_2-hazard.

3. The D_1-trio path is $00 \rightarrow 10 \rightarrow 11 \rightarrow 10$ under branching conditions $E\bar{C}\bar{M} \rightarrow EC\bar{M}$ due to an unintended delay of Δt_{D1} on the initiator C line to the ANDing RG $y_1\bar{C}$ (gate 4) in y_0. The IP must not be inconsistent with $\bar{y}_1, \bar{y}_0, E, \bar{M}$ and must contain the initiator as either C or $\bar{C}$ in y_1. Thus, the IP and the minimum path delay required to activate the D_1-trio are the same as those for E_1-hazard activation in item (1). This D_1-trio will cause a glitch in output P.

4. The D_2-trio path is $01 \rightarrow 00 \rightarrow 10 \rightarrow 00$ under branching conditions $ECM \rightarrow \bar{E}CM$ due to an unintended delay of Δt_{D1} on the initiator E line to the ANDing RG $EC\bar{X}$ (gate 2, which

contains $\bar{y}_0 EC$) in y_1. The IP must be via only gate 5 giving a minimum path delay requirement for the D_2-trio activation of $\Delta t_{E1} > \tau_5$, a relatively small delay. However, activation of the D_2-trio cannot cause a glitch in P but does cause a glitch in y_1.

It is clear that neither E-hazard nor the D_1-trio are a cause for concern because the minimum delay requirements to activate them are relatively large, all exceeding two gate delays. If there is a concern, however, counteracting delays of one or two τ_p placed on the second invariant feedback line will more than suffice to endure proper operation of the FSM.

7.3   EXAMPLE OF AN STT FSM ANALYSIS

Let us suppose we are given a logic circuit whose inputs to three complementary C-element modules and whose outputs are given by

$$\left\{\begin{array}{l} S_2 = \bar{A}B + \bar{y}_0 B \\ R_2 = A\bar{B} + y_0 A \\ S_1 = AB \\ R_1 = \bar{A}\bar{B} \\ S_0 = \bar{y}_1 \bar{A}B + y_1 A\bar{B} + y_2 \bar{A}\bar{B} \\ R_0 = \bar{y}_1 A\bar{B} + y_1 \bar{A}B + \bar{y}_2 \bar{A}\bar{B} \\ W = \bar{y}_2 y_1 y_0 \\ X = y_2 y_1 \bar{y}_0 A \\ Z = y_2 \bar{y}_1 y_0 \end{array}\right\} \qquad (7.2)$$

From the information in Eqs. (7.2) and following the analysis procedure outlined in Section 7.1, we will analyze this FSM ending with a detailed state diagram showing all states and branching paths associated with the FSM. Figure 7.5 shows the $SR \to Y$ K-map conversions for the NS logic. Note that the conversion process is just the reverse of the four-step $Y \to SR$ conversion algorithm given in Section 2.1. The reader should trace through this K-map conversion to verify the reverse algorithm.

From the LPD EV K-maps in Figure 7.5, we construct the PS/NS table shown in Figure 7.6a, where the external inputs (A and B) are listed in literal-canonical binary sequence for each of the eight present states 000 through 111. This PS/NS table is the tabular form of the state diagram presented in Figure 7.6b. Note that there are four primary states (shaded) and that each transition

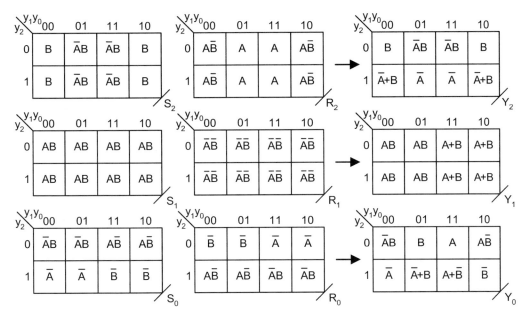

FIGURE 7.5: SR-to-Y K-map conversion for the NS functions given in Eqs. (7.2).

between any two of these states represents a Hamming distance of 2, as was true for the one-hot designs. There are four *don't care states* easily recognized as those having only out-branching conditions, and no in-branching paths. As can be seen, transitions between any two primary states must briefly transit via one of the don't care states but always with the appropriate branching conditions, again similar to the one-hot designs but faster. Note that this state diagram, with reference to the primary states, is the same as that shown in Figure 5.2, but this FSM is now identified as a single transition time (STT) machine. Thus, there are no cycles, buffer states, critical races, or output race glitches, and every state-to-state transition must pass through a don't care state.

A simulation of the FSM, represented by Eqs. (7.2) using C-elements as memory, is provided in Figure 7.7a. Note that it compares nearly identically with that in Figure 5.4, except that it has only three state variables instead of four as required for the one-hot design. Figure 7.7b shows a blowup of the shaded area in Figure 7.7a showing the 110 → 011 transition via a very brief passage through don't care state 111. It is true that both one-hot and STT designs of this FSM must transit between states with a Hamming distance of 2 (i.e., between states differing by two 1's in their state code assignment). However, it is the STT design that is the faster of the two designs as can be seen

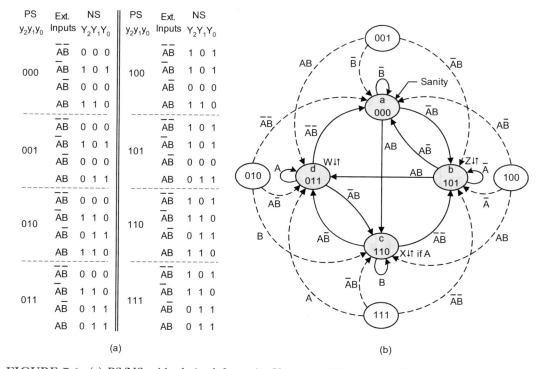

FIGURE 7.6: (a) PS/NS table derived from the K-maps of Figure 7.5. (b) State diagram, as constructed directly from the PS/NS table in (a), showing the four primary states (shaded) and the four don't care states for which there are no in-branching conditions, only out-branching paths.

by comparing the blowup regions in Figures 5.4b and 7.7b. In general, STT FSMs are the fastest machines possible.

A static hazard analysis indicates that there are two externally initiated static-1 hazards in NS forming logic given by Eqs. (7.2). One hazard exists in S_0 between coupled terms $\bar{y}_1 \bar{A} B$ and $y_2 \bar{A} \bar{B}$, and is produced on a $B \rightarrow \bar{B}$ change while in state 101 holding $\bar{A}$ constant. The cover for this hazard is $y_2 \bar{y}_1 \bar{A}$. The second static hazard exists in R_0 between coupled terms $\bar{y}_1 A \bar{B}$ and $\bar{y}_2 \bar{A} \bar{B}$, and is produced on $A \rightarrow \bar{A}$ change while in state 000 holding $\bar{B}$ constant. However, neither of these hazards can get by the C-elements and, consequently, no hazard cover is needed. This is one advantage in using C-element for the memory stage in asynchronous FSM design. However, there is always the possibility that function hazards will be formed if the inputs are allowed to change in close proximity to each other. Remember that a function hazard glitch is a correct response to competing input

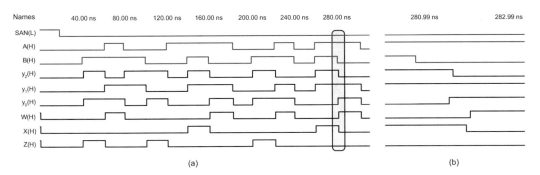

FIGURE 7.7: (a) Simulation of the STT FSM represented by Eqs. (7.2) implemented with C-elements and in agreement with Figure 7.6. (b) Blowup of the shaded area in (a) showing the 110-to-011 transition taking place via don't care state 111.

changes. To avoid function hazards, input changes should be minimally separated so as to avoid runt pulses that can cause any memory element to go metastable.

Because E-hazards are potential sequential defects, the minimum requirements for E-hazard formation are the same as those given in Section 5.4, but without the need to include a specific cycle state as was required for a one-hot E-hazard analysis. Thus, an E-hazard analysis will follow the same procedure as laid out in Section 3.6. The FSM in Figure 7.6b has two E-hazards that are possible. The first will occur along a path $a \rightarrow c \rightarrow d$ under input conditions $A\bar{B} \rightarrow AB$ with the inadvertent delay Δt_{E1} on the initiator line B to the first invariant y_0. The RG must be the ANDing RG $y_1 A\bar{B}$ in y_0 found in S_0, and the IP must not be inconsistent with the origin state $\bar{y}_2$, $\bar{y}_1, \bar{y}_0$ and the constant input A, and must contain the first invariant as B or $\bar{B}$ in y_1. Therefore, the IP must be via AB in S_1. From this information, the minimum path delay requirement must be $(\Delta t_{E1} + \tau_{Inv}) > (\tau_{AB} + \tau_{C\text{-element}})$, hence very similar to the requirement indicated in Section 5.4 for the E_1-hazard.

The E_2-hazard will occur along a path $b \rightarrow d \rightarrow c$ under branching conditions $\bar{A}B \rightarrow AB$ with an unintended delay Δt_{E2} on the A line to the first invariant y_0. The RG is the ANDing RG $y_1 \bar{A}B$ in y_0 found in R_0. The IP must not be inconsistent with the origin state $y_2, \bar{y}_1, y_0$ and the constant input B, and must contain the first invariant as A or $\bar{A}$ in y_1. Thus, the IP must be via AB in S_1. Accordingly, the minimum path delay requirement is $(\Delta t_{E2} + \tau_{Inv}) > (\tau_{AB} + \tau_{C\text{-element}})$.

Clearly, any given E-hazard analysis is remarkably the same irrespective of the method used to implement the asynchronous FSM. In the case of the analysis just completed, we see that the

RGs, IPs, and minimum path delay requirements for E-hazard formation were basically the same. This confirms the fact that E-hazards are sequential defects that are inherent in the sequential behavior, and not the implementation method or model used.

7.4 EXAMPLE OF A ONE-HOT FSM ANALYSIS

The logic read directly from the logic circuit of a one-hot LPD FSM is given by the NS and output forming logic in Eqs. (7.3). Note that there are six state variables, two inputs (X and Y), that the FSM is initialized into state a via a Sanity input to the 000000 state, a *1-hot + zero* configuration. The 1-hot + zero term in Y_a results from simplification by using the factoring and absorptive laws to give $a\bar{b} + \bar{a}\bar{b}\bar{c}\bar{d}\bar{e}\bar{f} = \bar{b}\bar{c}\bar{d}\bar{e}\bar{f}$. Also, there are two outputs, P and Q, representing both Mealy and Moore outputs. For convenience, the state equations are given using state identifiers rather than state variables. Note that the first term in each Y_i is the holding condition for that state, and that the last term in each Y_i (the penultimate term in Y_a) are the "out of" terms as required in Eqs. (5.1). Finally, note that there are no static hazards possible in this FSM. An S-hazard that exists in an NS function will be an internally initiated static-1 hazard between "into" and "out of" terms, and will be covered by the holding condition in that NS function. The apparent externally initiated static-1 hazard in the output function P between coupled terms $c\bar{X}Y$ and fXY cannot exist because the FSM cannot reside in two one-hot states (c and f) simultaneously.

$$
\begin{array}{ll}
& \overset{\displaystyle 1\text{ - Hot} + 0}{\overset{\displaystyle \text{Term}}{\overbrace{}}} \\
Y_a = a\bar{X} + f\bar{X}Y + a\bar{b} + \bar{b}\bar{c}\bar{d}\bar{e}\bar{f} & \text{(after simplification)} \\
Y_b = bX + aX + eX\bar{Y} + b\bar{c} \\
Y_c = c\bar{X} + b\bar{X} + \bar{c}\bar{e}f \\
Y_d = d\bar{X} + e\bar{X}\bar{Y} + d\bar{f} & (7.3) \\
Y_e = eY + cXY + e\bar{b}\bar{d} \\
Y_f = fX + f\bar{Y} + cX\bar{Y} + dX + f\bar{a} \\
P = b + c\bar{X}Y + d + fXY \\
Q = c + d\bar{X}Y
\end{array}
$$

Because the NS and output forming logic for a one-hot FSM can be read directly from a state diagram or state table, the reverse is also true. Shown in Figure 7.8 is the state diagram for this one-hot FSM as constructed directly from Eqs. (7.3). An inspection of the state diagram reveals that there are no buffer states, cycles, or critical races present. There are no static hazards possible, but function hazards are possible if the external inputs or state variables change in close proximity to

TABLE 7.1: Potential E-hazards in Figure 7.8 showing split paths to states f and e for E_1- and E_2-hazards and an extended state for E_3-hazard

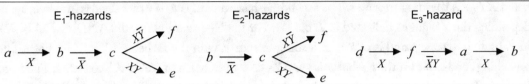

each other. It happens rarely but some function hazards cannot be avoided and must be dealt with by other means such as the use of C-elements as memory and, if necessary, output filters.

There are five potential E-hazards associated with the FSM state diagram in Figure 7.8. Referring to Section 5.4, the first E_1-hazard paths is $a \rightarrow b \rightarrow c$ that divides into two depending on

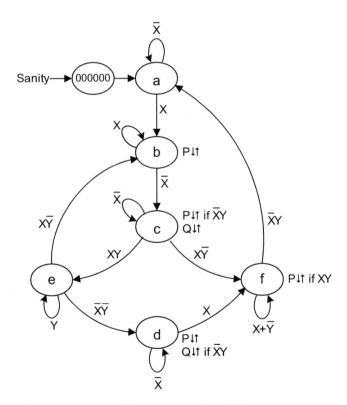

FIGURE 7.8: State diagram constructed directly from Eqs. (7.3) representing a one-hot asynchronous FSM.

the Y input as indicated in Table 7.1. For both E_1-hazard paths, the ANDing RG is $b\bar{X}$ in Y_c, and the IP is aX in Y_b. For these sequences, the branching conditions must be $\bar{X}\bar{Y} \rightarrow X\bar{Y}$ leading to state f, and $\bar{X}Y \rightarrow XY$ leading to state e. The second E_2-hazard path is $b \rightarrow c$ that divides into two, again depending the Y input as indicated in Table 7.1. For the E_2-hazard paths, the ANDing RG is $cX\bar{Y}$ in Y_f leading to state f, and is cXY in Y_e leading to state e, with the IP being $b\bar{X}$ in Y_c for both. The branching conditions for the E_2-hazards are $X\bar{Y} \rightarrow \bar{X}\bar{Y}$ to state f, and $XY \rightarrow \bar{X}Y$ to state e. The third E_3-hazard path is $d \rightarrow f \rightarrow a \rightarrow b$ under branching conditions $\bar{X}Y \rightarrow XY$ for which the RG is $f\bar{X}Y$ in Y_a with the IP being dX in Y_f. For all five E-hazards, the initiator is X but the minimum path delay requirements differ for them. To activate either the E_1-hazards or the E_3-hazard, the minimum path delay requirement is $(\Delta t_E + \tau_{Inv}) > (\tau_{IP} + \tau_{ORing})$, where the right side of the inequality is an ANDing operation into an ORing operation as in the SOP logic of Eqs. (7.3). To activate the E_2-hazards, the minimum path delay requirement is $(\Delta t_E) > (\tau_{Inv} + \tau_{IP} + \tau_{ORing})$. As a reminder, only a single change of the initiator is allowed in any $\alpha \rightarrow \beta \rightarrow \gamma$ E-hazard path. Thus, paths $c \rightarrow e \rightarrow d$ and $c \rightarrow f \rightarrow a$ cannot produce an E-hazard. Path $c \rightarrow e \rightarrow b$ is excluded as an E-hazard path because only one initiator change is permitted.

To test the sequential functionality of this FSM, excluding E-hazard production, a simulation of the logic circuit (circuit not shown) is provided in Figure 7.9. Here, all state-to-state transitions and output responses are shown. The logic circuit from which this simulation is made is NAND-

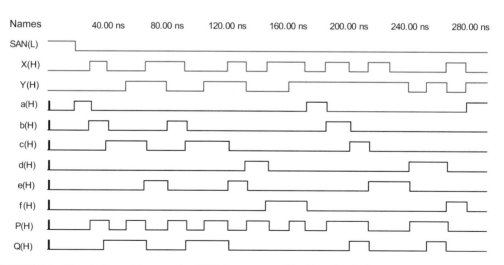

FIGURE 7.9: Simulation of the one-hot FSM represented by Eqs. (7.3) showing strict adherence to its state diagram in Figure 7.8.

based with C-elements and the appropriate inverters. Note that each one-hot state-to-state transition must transit through a state containing two 1's, one from the origin state and the other from the destination state.

7.5 EXAMPLE OF A PULSE MODE FSM ANALYSIS

Equations (7.4) contain the NS and output logic functions that are read from a pulse mode FSM having three-state variables, two inputs (X and Y) and two outputs, P and Q. The FSM is to be initialized into the 000 state. These functions are mapped in Figure 7.10 and converted to LPD form by using the $T \rightarrow Y$ K-map conversion algorithm. This algorithm is stated in Part 2 of Section 7.1.

$$T_2 = \bar{y}_2 \bar{y}_1 \bar{y}_0 Y + y_2 y_0 X$$
$$T_1 = \bar{y}_2 y_1 y_0 X + y_0 Y$$
$$T_0 = \bar{y}_2 \bar{y}_1 \bar{y}_0 X + y_2 \bar{y}_1 y_0 Y + y_1 (y_2 \oplus y_0)(X + Y) \qquad (7.4)$$
$$P = y_2 y_1 X + y_1 y_0 X + \bar{y}_2 \bar{y}_1 y_0$$
$$Q = y_2 y_1 Y$$

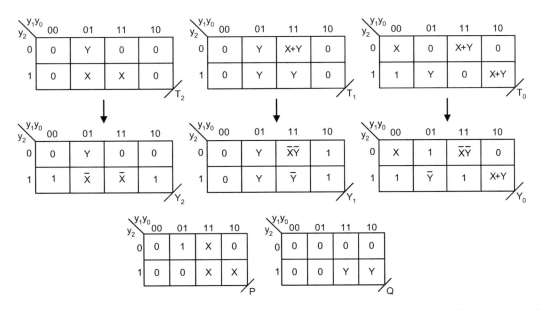

FIGURE 7.10: NS T-to-Y K-map conversion and output logic for the pulse mode FSM represented by Eqs. (7.4).

Now that we have mapped the NS and output logic into LPD EV K-maps, we can construct the PS/NS table. Construction of a PS/NS table for a pulse mode FSM is remarkably simple because the inputs must be presented as discrete nonoverlapping positive pulses that are at least minimally separated. This means that only input combinations such as $\alpha\bar{\beta}\bar{\gamma}, \bar{\alpha}\beta\bar{\gamma}$, or γ need be considered in the PS/NS table; all others, such as say $XY, \bar{X}\bar{Y}$ or $\bar{Y}$, must be discarded or ignored. Thus, only one input can be active at any given time, and a transition cannot occur on an input condition of "no active input." Shown in Table 7.2 is the PS/NS table as obtained directly from the NS EV K-maps in Figure 7.10. Remember that any given output is given with respect to the PS, never with respect to the NS.

Construction of the state diagram for this pulse mode FSM is accomplished directly from the PS/NS table in Table 7.2 and is shown in Figure 7.11. Here, input entries such as $X\bar{Y}, \bar{X}Y, X$, or Y simply mean that an X or Y pulse is required to execute a given transition on the falling edge of that pulse. States 010 and 100 are hang states, meaning that should the FSM power up into either of these states the FSM would reside in that state indefinitely. The dashed arrow from state 100 to 101 is used to indicate that the PS/NS table calls for that transition but which is clearly not possible due to the absence of any single, nonoverlapping positive pulse. The presence of hang states addresses the importance of initializing into a specific state such as the 000 state. To do otherwise would be a serious mistake and could lead to unrecoverable errors.

TABLE 7.2: PS/NS table as constructed directly from the EV K-maps in Figure 7.10 representing the pulse mode FSM given in Eqs. (7.4)

PS		NS			PS		NS		
$y_2 y_1 y_0$	INPUTS	$Y_2 Y_1 Y_0$	P	Q	$y_2 y_1 y_0$	INPUTS	$Y_2 Y_1 Y_0$	P	Q
000	X	001	0	0	100	$-$	101	0	0
001	Y	111	1	0	101	$\bar{X}Y$	110	0	0
						$X\bar{Y}$	001		
010	$-$	010	0	0	110	$\bar{X}Y$	111	X	Y
						$X\bar{Y}$	111		
011	$\bar{X}Y$	000	X	0	111	$\bar{X}Y$	101	X	Y
	$X\bar{Y}$	000				$X\bar{Y}$	011		

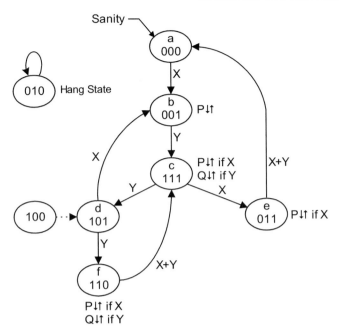

FIGURE 7.11: State diagram obtained from the PS/NS table in Table 7.2 showing extraneous states 010 and 100.

To complete the analysis of this pulse mode FSM, we iterate the benefits of this approach to FSM design. As with all properly designed pulse mode FSMs, the state machine in Figure 7.11 cannot have endless cycles, critical races, or static hazards in either the NS or output logic, output race glitches, and E-hazards. All these benefits result from the stringent requirements placed on the inputs to any given pulse mode FSM. That is, all inputs must be presented as discrete nonoverlapping positive pulses that are at lease minimally separated and that all state transitions occur on the falling edge of an input pulse. To meet these input restrictions, it may be necessary to use bus arbiters on the incoming data, a subject to be discussed in Chapter 11 and covered by Problem 11.2 in Appendix B.

PART II

Self-Timed Systems, Programmable Sequencers, and Arbiters

CHAPTER 8

Externally Asynchronous/Internally Clocked Systems

Externally asynchronous/internally clocked (EAIC) systems operate somewhere between synchronous and asynchronous design methodologies. To the external world, EAIC systems are entirely asynchronous—there is no externally supplied system clock. Internally, the EAIC system is controlled by a clock signal that is generated following rendezvous, at the clock generation circuitry, of *all* "data ready" signals issued from the memory modules. Thus, if a data ready signal is delayed for any reason, the system pauses until that data ready signal has successfully rendezvoused with all other such signals at the clock generation circuitry—a *pausable* system. The speed of the clock operates as fast as the EAIC logic permits and can reach into the high MHz range. Properly designed memory modules can lead to extremely large *mean time between failures* (MTBF) virtually independent of the internal clock frequency. Furthermore, the internal clocking of the EAIC system guarantees the absence of endless cycles, critical races, essential hazards, and errors caused by static, dynamic and function hazards. Finally, there is the issue of metastability, a subject that will be effectively dealt with later in this chapter.

8.1 BASIC ARCHITECTURE AND SYSTEM CHARACTERISTICS

Shown in Figure 8.1 is the block diagram representing the generalized architecture for the EAIC system. It consists of input and memory registers, each containing DFLOP memory modules (similar to D flip-flops), next state (NS) and output forming logic, and clock (CK) generating circuitry, which is simply a specially built NOR gate. A tri-state driver is added to the output of the NOR gate so that the clock signal can be turned off at any time, forcing the EAIC system into a standby condition. Notice that a data ready signal, R, must be issued $0(H) = 1(L)$ by each DFLOP, and that all such signals must rendezvous at the NOR gate before a CK signal is generated $1(H)$. Thus, the CK signal is issued active high to each of the memory DFLOP modules but only when all data ready signals have arrived active low at the NOR gate (the conjugate form). At this time, data are clocked into the input register and previously clocked-in data are clocked out of the memory register. Then,

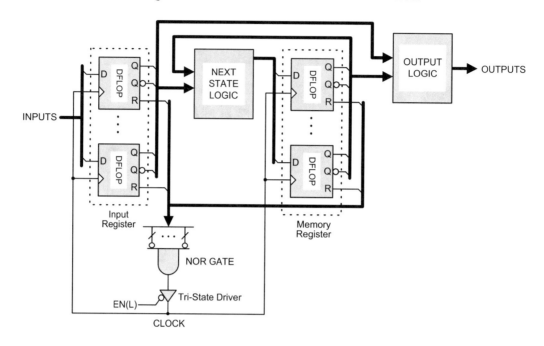

FIGURE 8.1: General Mealy model architecture for the EAIC system showing DFLOP input and memory registers and clock generating circuitry with tri-state enable.

when the first DFLOP senses a resolved set or resolved reset, CK is turned off, $CK(H) \rightarrow 0(H)$, by the active data ready signal $R_j \rightarrow 1(H) = 0(L)$. At this point, the input register is made ready to receive the next set of input data—a two-step process that provides numerous advantages over both synchronous and asynchronous systems. In the following, we will show that EAIC systems possess a number of features that make them an attractive choice for system-level applications. These features include controllers for which the internal clock is used to operate numerous data-path devices such as counters, shift registers, etc., and the creation of a highly reliable delay-insensitive or pausable mode of operation that creates a near-infinite MTBF. In most cases, input conditioning circuits such as arbiters and synchronizers are not needed but debouncers may be required. The use of de-bouncing circuits is not as critical as in the case of pulse mode designs, where debouncing circuits should be considered mandatory (see Section 6.5).

The logic circuit for a multiple input NOR gate, provided in Figure 8.2a, consists of a special p-channel MOSFET (PMOS) and a bank of n-channel MOSFET (NMOS) such that CK(H) can go active only if all data ready inputs go to LV, that is all $R_i(H) \rightarrow 0(H) = 1(L)$. This NOR gate is specifically designed to minimize *fan-in* limitations and propagation delay. The number of permissible inputs R_i up to about eight will have negligible effect on the gate path delay, a condition not

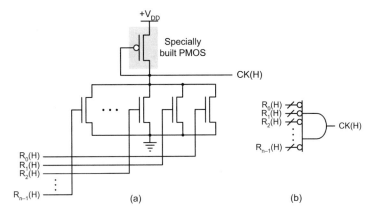

FIGURE 8.2: Multiple input NOR gate specifically designed to minimize fan-in limitations and propagation delay. (a) CMOS circuit. (b) Generalized NOR gate symbol and input logic level requirements for EAIC system operation.

true for a normal complementary metal-oxide conductor (CMOS) NOR gate. To work properly, the special PMOS must be designed such that the drain-to-source impedance remains sufficiently high as to minimize drain current when one or more NMOS are turned ON. In this respect, the special PMOS operates similarly to a depletion-mode NMOS. The generalized NOR symbol is shown in Figure 8.2b.

8.2   DFLOP MEMORY ELEMENT DESIGN WITH C-ELEMENTS

Shown in Figure 8.3 are the block diagrams and state diagrams for a DFLOP memory element appropriate for use in the EAIC system depicted in Figure 8.1. It consists of a C-element resolver (a), a C-element output memory stage, and a NAND gate required to generate the data ready signal, R. Below the block diagrams are the state diagrams. The state diagram for the resolver is the same as that used for the design of an RET (rising edge triggering) D flip-flop as discussed in detail in Tinder's text (see Endnotes). The output memory stage is that for a complementary C-element as shown in Figure 1.13 and needs no further discussion. The resolver, on the other hand, requires considerable attention to produce the best resolver unit possible, one that will serve to produce the desired characteristics mentioned at the beginning of this chapter.

To better understand how the EAIC system in Figure 8.1 operates, consider a blowup of the CK waveform segment in Figure 8.4 produced by the CK generating NOR gate. After initializing into the 00 state of the resolver, the CK signal is issued at a frequency dependent only on the speed of the EAIC logic but independent of the data activation levels to the input registers. The internal clock CK goes to $1(H)$ only when *all* data ready $R(H)$ signals go to $0(H) = 1(L)$, and *all* DFLOP resolvers are in their unresolved 00 state. Any delay of one or more DFLOPs causes CK activation

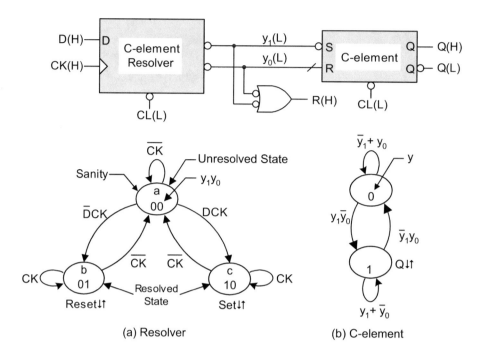

FIGURE 8.3: Design of the C-element-based DFLOP for EAIC systems and showing the NAND gate required to generate the data-ready signel. (a) State diagram for the C-element resolver FSM input stage. (b) State diagram for the complementary C-element output stage with output, Q.

to pause until *all* data ready signals rendezvous properly at the NOR gate. Thus, $CK = 1(H)$ only when all $y_j(L) = 0(L)$. At this time, external data are clocked into the input registers and previous data are clocked out of the memory registers. However, the first data ready $R_j(H)$ signal that senses a resolved Set or resolved Reset of a DFLOP's $y_1(L), y_0(L)$ state variables sends a $1(H) = 0(L)$ to the

Rising Edge
(RE)

Falling Edge
(FE)

RE: $CK \rightarrow 1(H)$ when *all* R's $\rightarrow 0(H) = 1(L)$. That is, when $y_1, y_0 \rightarrow 10$ (Set) or 01 (Reset)

FE: $CK \rightarrow 0(H)$ when *first* $R_j \rightarrow 1(H) = 0(L)$ and *all* $y_1(L) = y_0(L) = 0(L)$ in unresolved state 00.

FIGURE 8.4: Blowup of the CK waveform segment showing requirements for the continuous generation of the internal clock in the EAIC system.

CK generating NOR gate. This, in turn, causes CK to go to $0(H)$ and all resolvers to transit to their unresolved 00 states ready to begin the process all over again.

The entered variable (EV) K-maps for the design of the C-element-based DFLOP resolver are given in Figure 8.5. By using the mapping algorithm in Section 1.6, the lumped path delay (LPD) K-maps are produced from Figure 8.3a as shown in Figure 8.5a. They are then converted to SR K-maps in Figure 8.5b by using the $Y \rightarrow SR$ conversion algorithm described in Section 2.1. From these EV K-maps, the optimum logic expressions are easily extracted and are presented in Eqs. (8.1) and (8.2). The logic expressions for the LPD (Huffman) DFLOP design are given in Eqs. (8.1), which when implemented operates in the fundamental mode. Equations (8.2) give the logic expressions for the Muller C-element DFLOP design. When these expressions are implemented, the DFLOP will operate as a quasi Muller circuit because C-elements operate outside of the fundamental mode. It is the C-element design we emphasize in this text because they provide the most protection and the highest reliability. There are, however, two other ways to design the DFLOP—by using basic cells and by using a different LPD design (see Tinder's text in Endnotes)—but both must operate in the fundamental mode. Such designs do not afford the same level of protection and reliability as the C-element design, but are otherwise completely acceptable.

$$\left\{\begin{array}{l} Y_1 = \bar{y}_0 DCK + y_1\bar{y}_0 CK = (D + y_1) \cdot \bar{y}_0 CK \\ Y_0 = \bar{y}_1 \bar{D}CK + \bar{y}_1 y_0 CK = (\bar{D} + y_0) \cdot \bar{y}_1 CK \end{array}\right\} \text{Huffman DFLOP resolver} \qquad (8.1)$$

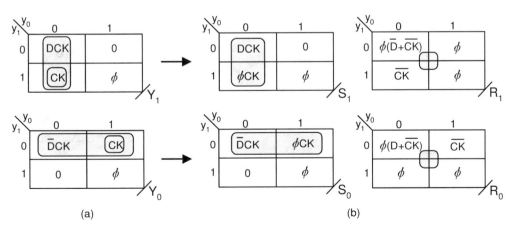

(a)   (b)

FIGURE 8.5: EV K-maps for the resolver input stage in Figure 8.3a. (a) LPD K-maps obtained by using the mapping algorithm given in Section 1.6. (b) SR K-maps obtained by using the Y-to-SR conversion algorithm in Section 2.1 for use in the C-element based resolver of a DFLOP.

$$\left.\begin{cases} S_1 = \bar{y}_0 DCK & \bar{R}_1 = CK \\ S_0 = \bar{y}_1 \bar{D}CK & \bar{R}_0 = CK \end{cases}\right\} \text{Quasi-Muller DFLOP resolver} \tag{8.2}$$

8.2.1 D-Trio Analysis of the Resolver FSM

There are two d-trios in the resolver circuit of Figure 8.3a. Taking the initiation state as 01, the first d-trio D_1 path is $01 \rightarrow 00 \rightarrow 10 \rightarrow 00$ if $DCK \rightarrow \bar{D}\bar{C}K$ occurs in state 01 when an unintended delay Δt_{D1} of sufficient magnitude is placed on the CK line to an ANDing race gate (RG_1). From the state diagram the ANDing RG is easily identified as $\bar{y}_0 DCK$ in the first invariant Y_1, or for a quasi-Muller resolver and use of complementary C-elements the RG is $\bar{y}_0 DCK$ in S_1. Following Section 3.6.1, the indirect path must be via gate $\bar{y}_1 \bar{D}CK$ in S_0. From this information, it is clear that the minimum requirements for d-trio formation would be $\Delta t_{D1} > (\tau_{NAND} + \tau_{C-element} + \tau_{Inv})$. If each gate and C-element is assigned 0.3 ns with inverters 0.1 ns, then $\Delta t_{D1} > 0.7$ ns. An active D_1 would glitch the output Set⬇⬆. It is left as an exercise for the reader to analyze the second d-trio D_2. Hint: It will be very similar to that for the first d-trio just described.

The logic circuit for the resolver, as derived from the expressions in Eqs. (8.2), is shown in Figure 8.6a. Note that two C-elements are required for the resolver stage and one for the output stage. Also, note that $R_1 = R_0 = \bar{CK}$, meaning that when used with a complementary C-element $\bar{R} = CK$, as shown in Figure 8.6a. The two inverters on the $y_i(L)$ lines can be removed by reversing the inputs to the output C-element and by replacing the NAND gate by an OR gate. Such a change

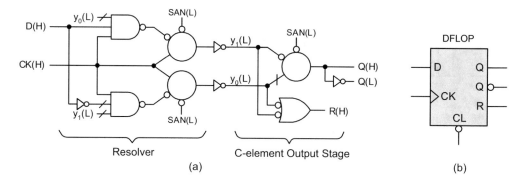

FIGURE 8.6: C-element design of a Muller DFLOP. (a) Logic circuit as designed from Eqs. (8.2) and Figure 8.3 showing the resolver logic, output stage, and the data ready NAND gate. (b) Circuit symbol for the DFLOP in agreement with (a).

presents only minor or insignificant changes in the operation of the DFLOP. The logic circuit symbol for the DFLOP is given in Figure 8.6b.

8.3   SIMPLE EXAMPLE OF AN EAIC FSM DESIGN

We illustrate the application of the EAIC system by designing the Gray-to-Seven sequence recognizer whose state diagram is presented in Figure 8.7a. Here, the Gray code sequence is presented to the finite state machine (FSM) as inputs X, Y, and Z and is recognized only when the input sequence reaches the binary seven, XYZ in state 111 issued as a pulse *Seq_Rec*.

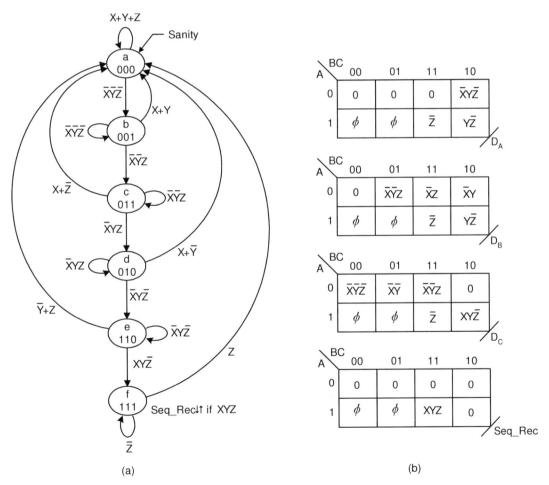

(a)   (b)

FIGURE 8.7: (a) State diagram for a three-bit Gray-to-Seven FSM with a single output. (b) EV K-maps for the three state variables and output Seq_Rec.

The EV K-maps for the NS and output logic are given in Figure 8.7b as plotted directly from the state diagram in Figure 8.7a by applying the mapping algorithm given in Section 1.6 together with the D exitation table given in Figure 1.2. The advantage of the EV mapping approach can be appreciated by the fact that use of the old conventional 1's and 0's method would require the use of sixth-order K-maps. Even so, the third-order EV K-maps in Figure 8.7b are sufficiently complex enough to justify the use of a logic minimizer that accepts EVs. The recommended logic minimizer is called BOOZER, and its results are given by Eqs. (8.3), (8.4), and (8.5). As is evident, BOOZER seeks out and maximizes the use of shared PIs thereby providing optimum or near optimum results. (See Preface for a brief description of all recommended software for use with this text.)

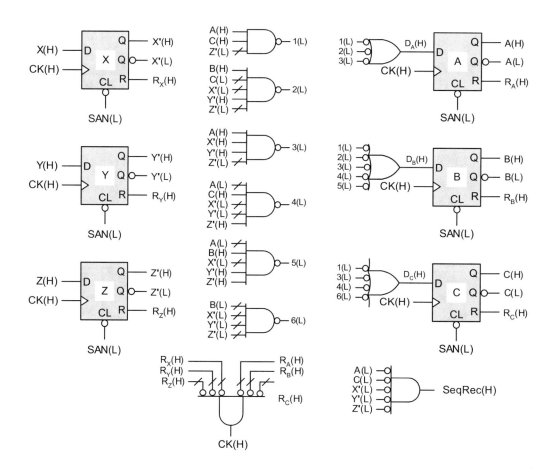

FIGURE 8.8: Logic circuit for the Gray-to-Seven FSM constructed from Eqs. (8.3), (8.4), and (8.5) by using the wireless connection feature and in agreement with the EAIC architecture shown in Figure 8.1.

$$[1] = AC\bar{Z} \ [2] = B\bar{C}\bar{X}Y\bar{Z} \ [3] = AXY\bar{Z} \ [4] = \bar{A}C\bar{X}\bar{Y}Z \ [5] = \bar{A}B\bar{X}YZ \ [6] = \bar{B}\bar{X}\bar{Y}Z \quad (8.3)$$

$$D_A = [1] + [2] + [3]; \quad D_B = D_A + [4] + [5]; \quad D_C = [1] + [3] + [4] + [6] \quad (8.4)$$

$$\text{Seq\_Rec} = ACXYZ \quad (8.5)$$

The logic expressions for the Gray-to-Seven sequence recognizer given by Eqs. (8.3), (8.4), and (8.5) are shown implemented in Figure 8.8, where the mixed-logic wireless connection feature (emphasized in this text) is used to simplify the appearance of the logic circuit. Here, use is made of the C-element DFLOP circuit symbol given in Figure 8.6b and the clock generating circuit is a six-input NOR gate but with no tri-state driver on its output.

Simulation of this logic circuit is given in Figure 8.9 showing an internal clock frequency of about 417 MHz, the Seq_Req(H) output pulse in state 111, and a static-1 hazard in the D_B memory register input that could not get passed the C-elements in its DFLOP. Note that the FSM is initialized into the 000 state, consistent with the state diagram in Figure 8.7a and the logic circuit in Figure 8.8. When the sanity input releases the FSM to operate normally, the FSM begins in state

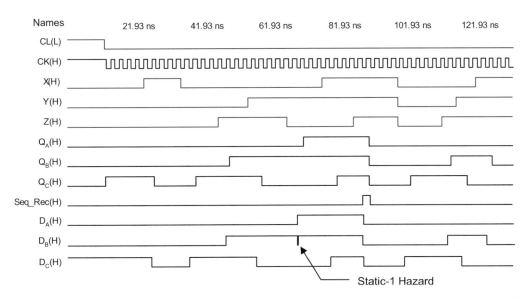

FIGURE 8.9: Simulation of the EAIC circuit in Figure 8.7 showing the internal clock signal, the NS and output response to input changes, and a static-1 hazard in the input to an output stage DFLOP.

001 as required by the state diagram when the input conditions are $\bar{X}\bar{Y}\bar{Z}$. To initialize this FSM reliably and stably into state 000, at least one of the inputs must be active at the time the sanity input signal goes inactive. Note that the static-1 hazard in D_B is externally initiated in the 010 state under holding conditions $\bar{X}YZ$ when $Z \to \bar{Z}$ as indicated in Eq. (8.6). From an inspection of Eq. (8.6), it is clear that the hazard cover for the static hazard in D_B would be $\bar{A}B\bar{C}\bar{X}Y$, but it is not needed because the C-element filters out the hazard.

$$D_B = AC\bar{Z} + B\bar{C}\,\bar{X}Y\bar{Z} + AXY\bar{Z} + \bar{A}C\bar{X}\bar{Y}Z + \bar{A}B\bar{X}YZ \qquad (8.6)$$

$$\underset{010}{\uparrow} \qquad\qquad \underset{\bar{X}Y}{\rule{8em}{0.4pt}} \qquad \underset{010}{}$$

8.4 THE METASTABLE DETECTION STAGE

We must acknowledge that no logic device is completely resistant to metastability. However, we know that C-elements operate outside of the fundamental mode, which makes them less susceptible to the metastable state. But under certain conditions, even C-elements can become metastable. Should such a state manifest itself in a DFLOP memory module, the *pausible* character of the EAIC system will allow the system to operate properly provided the DFLOP can internally exit from the metastable state as a clean set or reset. To ensure that the DFLOP module will perform this important function, it is necessary to use a special *metastable detection stage* (MDS) within the DFLOP. Such an MDS stage is shown in Figure 8.10 interposed between the resolver and output stage of the DFLOP. It is the function of the MDS to prevent any metastable condition produced within the resolver stage from being passed on to the output C-element stage. By adjusting the switching thresholds of the MDS gates down ($\downarrow$) and the inverters up ($\uparrow$), the overall *switching threshold* V_{th} is lowered to about a quarter of the supply voltage, or $V_{\text{th}} \cong 0.25V_{\text{DD}}$. This is accomplished by adjusting the PMOS/NMOS width ratios W_p/W_n of the MOSFETS to 1/4 for down adjustment ($\downarrow$) and

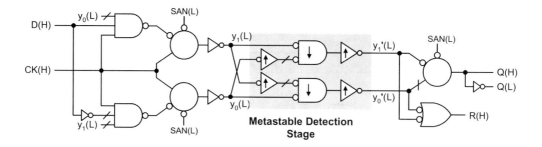

FIGURE 8.10: C-element design of a Muller DFLOP with a metastable detection stage (MDS) designed to prevent any possible metastable condition from being passed on to the output C-element and data-ready gate.

8/1 for the up adjustment ($\uparrow$). With the overall switching threshold altered down to about $0.25V_{DD}$, only strong active low input signals $y_1 < V_{th}$ will carry to the output y_1'. On the other hand, active low input signals $y_1 > V_{th}$ will not cross the switching threshold and will not carry to the output y_1'. As a result, only cleanly asserted active low signals can pass through the MDS circuit, whereas any metastable condition will cause the MDS to drop low after exiting the metastable state. The MDS stage also serves as a *mutual exclusion* operation. To better understand the above discussion, refer to Eqs. (1.8), where in mixed logic symbology it is clear that $0(H) = 1(L) \rightarrow LV$. Thus, only a strong active low $y_1(L)$ signal will correspond to $y_1 < V_{th}$ (i.e., LV) and be passed through to the output y_1'. A weak active low $y_1(L)$ signal for which $y_1 > V_{th}$ (or HV) would not cross the switching threshold (toward zero voltage) and would not carry to the output y_1'. Figure 8.11 illustrates these facts.

To test the action of the MDS circuit, its PSPICE simulation is shown in Figure 8.11, where response to a metastable condition, region (2) is featured after a brief period (1) of correct operation. Here, the metastable voltage V_m tends to lie in the usual range of mid-supply ($V_m \approx V_{DD}/2$). This is the voltage at which a metastable condition resides for a time (the *metastable exit time*, Δt_m) after which it must resolve as either a clean set or reset. Experiments have shown that the metastable exit time is highly unpredictable. Furthermore, there is never a guarantee that the metastable state will flatline on V_m for a time Δt_m. To test the MDS under worse-case conditions, the inputs, y_1 and y_0 in region (2) are introduced as a damped sign-wave oscillation about the threshold voltage V_{th} with

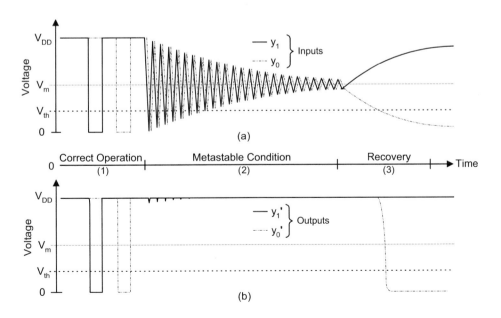

FIGURE 8.11: PSPICE simulation of the MDS circuit Figure 8.9b showing (a) input voltages and (b) output voltages before, during, and after a metastable state.

a phase difference of 90°. During this metastable state, the oscillatory condition causes only the beginning of pulse formation of the MDS outputs y_1' and y_0'. After a time Δt_m, the inputs resolve as either a clean set or reset causing the MDS outputs to drop predictably low after the inputs exceed the adjusted *threshold voltage* V_{th}. This means that the outputs of the MDS drop to $0(L)$ with a data ready output of $R(H) = 0(H)$ and a DFLOP output $Q(H) = 0(H)$. Under these conditions, it is fair to say the EAIC system would have an MTBF that is extremely high but not infinite.

8.5 FREQUENCY CHARACTERISTICS AND NS LOGIC CONSTRAINTS OF EAIC SYSTEMS

Tracing the CK cycle path in Figure 8.1 beginning at the input of the CK generating NOR gate and ending at its output, it is clear that the CK frequency must be given by

$$f_{CK} = (2\delta_{DFLOP} + 2\delta_{NOR})^{-1} \tag{8.7}$$

where δ_{DFLOP} is the propagation delay through the DFLOP and δ_{NOR} is the propagation delay through the NOR gate exclusive of the tri-state driver. If use is made of modern high-speed CMOS, as is assumed in this text, the CK frequency can exceed 400 MHZ as was demonstrated in Figure 8.9. Here, we assume that all gates and inverters have a propagation delay of 0.30 and 0.10 ns, respectively. For simplicity, no account is taken of gate fan-in although the simulator we use has that capability. Also, for the complementary C-elements we have assigned a propagation delay of 0.20 ns. Thus, a calculation made on this basis by using Eq. (8.7) gives a frequency of 417 MHz, which is exactly that measured in a blowup of Figure 8.9. If use is made of the MDS stage in Figure 8.10, the CK frequency would drop to 345 MHz.

If the path delay through the NS forming logic in Figure 8.1 exceeds a certain upper bound, the proper operation of the EAIC system cannot be guaranteed. To stay within this upper bound, it is necessary that the updated Q outputs from the input register propagate through the NS forming logic before the next rising edge clock event, that is, during one CK cycle. Tracing one CK cycle beginning and ending at the NOR gate, the NS logic constraint is given by

$$\delta_{NS} \geq (\delta_{DFLOP} + 2\delta_{NOR}) \tag{8.8}$$

which is more than four gate delays without the MDS or about six gate delays with the MDS. Thus, two-level logic with inverters is well within this upper bound and the proper operation of the EAIC system is guaranteed. Clearly, the minimum width of any data pulse must be larger than one CK cycle if it is to be picked up reliably by the EAIC system. The *throughput* is the elapsed time between input change and output response normally occurring in the range $\delta_{Throughput} = (3\delta_{DFLOP} + 2\delta_{NOR}) \pm f_{CK}^{-1}$.

8.6   PARALLEL/SERIAL PROCESSING WITH CASCADED EAIC MICROCONTROLLERS

8.6.1   Characteristics

EAIC microcontrollers (MCs) have all the advantages of synchronous designs but without an external clock. The internal clock operates as fast as the internal logic permits, even in the high MHz range. *EAIC MCs* are *delay-insensitive, pausable systems* with a *near-infinite MTBF*. The internal clock is not precisely regular but can operate any number of peripherals just as synchronous *MCs* can do. Logic noise and metastable conditions do not pass through the input and memory registers due to the action of C-elements and MDS stages in the DFLOPs. Thus, memory and input registers of the EAIC architecture are both "metastability hardened" due to the presence of MDS stages and C-elements that operate outside the fundamental mode. It is much more difficult to force a C-element into a metastable state than in other memory elements. Synchronizer stages and arbiters are not required on input signals to *MCs* even when such inputs compete for access—again the advantage afforded the C-elements in memory stages. The internal clock signal of an *EAIC MC* can be turned off (paused) any time an *MC* is not in use thereby reducing power dissipation. CAD software used to optimize D flip-flop designs can be applied to EAIC MCs. Conversion of the EAIC memory

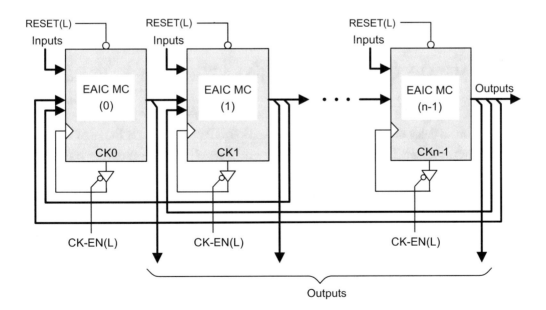

FIGURE 8.12: Cascaded configuration of *n*-EAIC microcontoller (MC) stages showing possible feedback variations between stages, external inputs to each stage, serial and parallel outputs, individual or global initialization (RESET), and individual or global internal clock enable (CK-EN).

elements (DFLOPs) to T and SR memory elements is accomplished via the usual means (see Tinder's text in Endnotes). Thus, counters, shift registers, and other state machine designs can be designed by either the EAIC method or by conventional means.

Parallel/serial processing can be achieved by cascading n-EAIC MCs, as that shown in Figure 8.12. Here, each MC stage can receive inputs from previous stages, from succeeding stages as feedback, or from external sources, and outputs can be generated from each stage. Each MC stage operates at its own internally generated clock frequency, which can be paused separately at any time or paused together with all or any combination of the remaining MCs. The major benefit of such a scheme is that it is programmable, highly reliable, and offers versatility not available by other schemes. When input data are idle, the internal clock of the MC stages can be turned off individually or globally, thereby minimizing power dissipation. Initialization (RESET) of the EAIC MC stages can also be done individually or globally. Finally, and most importantly, each MC stage shown in Figure 8.12 can be replaced by an MC together with its data path devices to produce a series/parallel system of minimicroprocessor (EAIC MM) stages that are fully programmable and that can handle complex operations with an extremely high degree of reliability and high speed. In all cases, it must be remembered that in a cascaded EAIC system of MCs, there is no single system clock that is common to all MCs. Each EAIC MC has its own individual internal clock that will normally not be in synch with those produced by other EAIC MCs. However, EAIC designs with MDS capability ensure the proper operation of a cascaded EAIC MC system with or without EAIC operated data path devices. Thus, clock skew problems become a non-issue.

8.7   SUMMARY OF THE SALIENT FEATURES OF EAIC SYSTEMS

1. EAIC systems are essentially delay-insensitive, meaning that the internal clock will pause anytime a delay occurs in the input register DFLOPs, or in the DFLOPs of the memory (output) register. The internal clock can be deliberately turned off during stand-by periods.

2. The internal clock can be used to operate any number of data path devices such as counters, shift registers, and other secondary state machines.

3. EAIC systems are metastably hardened, which raises the MTBF to extremely high values.

4. The CAD software ADAM can generate D functions for the design of FSMs suitable for use by with EAIC systems.

CHAPTER 9

Cascadable Asynchronous Programmable Sequencers (CAPS) and Time-Shared System Design

In Part I of this text, it was made clear that any finite state machine (FSM) designed to operate in the fundamental mode must be free of timing defects such as endless cycles, critical races, static hazards in the next state (NS) forming logic, and active essential hazards. Normally, the task of ridding the FSM of these defects is not difficult but it can be tedious and does require a considerable understanding of the intricacies of asynchronous FSM design methods. The externally asynchronous/internally clocked (EAIC) system, presented in Chapter 8, provides a means of avoiding these problems by using an internally generated clock somewhat similar to synchronous FSM design. However, the EAIC system cannot be used as a programmable sequencer owing to the mechanism required to generate the internal clock. In this chapter, we will consider in detail a versatile and highly reliable class of defect-free asynchronous programmable sequencers (APS) that can be cascaded to form sequencers of much greater capability—all defect-free. These APS can rightfully be classified as true Muller-type systems as defined in Section 1.10.

9.1 MICROPROGRAMMABLE ASYNCHRONOUS CONTROLLER MODULES

Simply put, the downfall of asynchronous designs is the inability of the designer to understand and deal with a variety of timing defects that can cause state machine failure. The classical approach to asynchronous system design requires that each design be carefully analyzed for a number of timing defects that may exist and then take whatever corrective action is necessary to eliminate them. This is *not* necessary for a PLD- or RAM-driven cascadable *APS* system approach as introduced here because such safeguards have been built into the programmable sequencers. The use of *microprogrammable asynchronous controller* (MAC) modules takes the guesswork out of the design process, making it possible for designers less knowledgeable in the intricacies of asynchronous design methods to produce reliable clockless APS systems. Cascadable MAC module APS systems can be operated under a variety of operating conditions and constraints, and on a time-shared basis, all defect-free, including E-hazards—time-, effort-, and expense-saving features.

9.2 MAC MODULE CHARACTERISTICS FOR USE WITH CAPS SYSTEM ARCHITECTURE

The following is a brief summary describing the salient features of the CAPS systems centered around MAC modules:

1. MAC modules can be driven by a bank of PLDs (see Glossary) or by RAM. MAC module APS systems can be instantly switched between radically different asynchronous micro-controllers (AMCs) on a time-shared basis, all defect-free. A *deactivate input* (*DI*) signal within the *MAC module* is used for extensive cycle control but can also be used for connectivity purposes making this APS system highly versatile. A DI signal pulse marks each state transition that occurs. The number of individual DI outputs needed depends on the number and size of cycle-state FSMs.

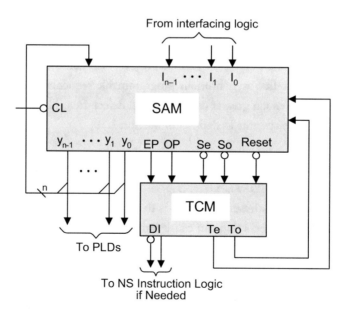

FIGURE 9.1: Components of an n-input MAC module consisting of a 2^n state array machine (SAM) and a timing control machine (TCM) and their interconnections.

2. Programming a MAC module is easily accomplished manually from a state diagram or from a state table and is, therefore, amenable to computer automated design, particularly for very large systems. The MAC module architecture has been published and has been tested in real time by using a field programmable gate array (FPGA) (see Tinder, Klaus, and Snodderley in Endnotes). A variety of state machines have been simulated by using MAC modules with Muller C-elements as memory. The simulation results for these machines show them to be very fast, highly reliable, and all defect-free. E-hazards are not possible.

3. Each MAC module consists of two asynchronous FSMs: a state array machine (SAM) and a timing control machine (TCM) in a *handshake configuration* as shown in Figure 9.1. A single n-bit *MAC module* can be used to produce any state machine of up to 2^n states with n-way branching capability. Cascading of n-bit modules is an attractive means of

TABLE 9.1: Various cascading configurations for MAC modules showing maximum state capacity, maximum number of state variables, and out-branching capability

BASE MODULE $(n \times m)$	NUMBER CASCADED	MAXIMUM STATE CAPACITY	MAXIMUM NUMBER OF STATE VARIABLES	OUT-BRANCHING CAPABILITY
2×2	1	$4 = (2^2)^1$	2	2
2×2	2	$16 = (2^2)^2$	4	4
2×2	3	$64 = (2^2)^3$	6	6
2×2	4	$256 = (2^2)^4$	8	8
2×4	1	$8 = (2)^{(1+2)}$	3	3
4×4	1	$16 = (4^2)^1$	4	4
4×4	2	$256 = (4^2)^2$	8	8
4×4	3	$4096 = (4^2)^3$	12	12
$2^l \times 2^m \times 2^n \ldots$	–	$2^{(l+m+n\ldots)}$	$l + m + n \ldots$	$l + m + n \ldots$

producing MAC module APS systems with very large state number capacities all without loss of speed or reliability.

4. MAC modules can be cascaded to produce very large APS systems all without compromising speed and reliability. For example, cascading three 2^n MAC modules produces a MAC module of 2^{3n} state capability and $3n$-way out-branching capability. Table 9.1 illustrates a few examples of MAC module cascading configurations

5. MAC module APSs require *logically adjacent* state-to-state transitions. Use of cycles and buffer states is permitted and is sometimes necessary. Each cycle or buffer state transition is strictly controlled by the handshake configuration between the SAM and TCM.

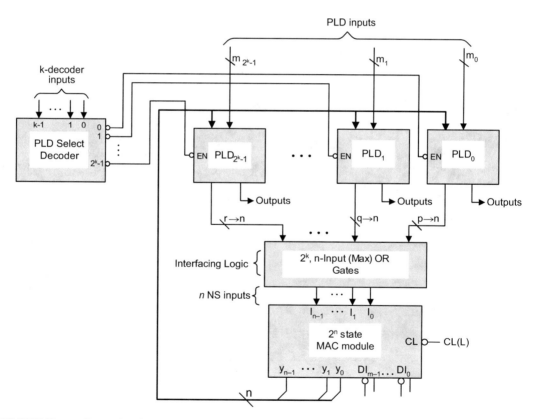

FIGURE 9.2: Generalized architecture for programming an n-bit MAC module to operate with any one of 2^k PLDs (ROMs, PLAs, PALs, or CPLDs) by using k-inputs to a PLD select decoder and by using 2^k, n-inputs to the interfacing logic.

6. The general architecture for programming an *n*-bit MAC module to operate as any one of 2^k asynchronous state machines or microcontrollers (AMCs) on a time-shared basis is shown in Figure 9.2. Here, the driving logic can be any number or combination of PLDs such as PLAs, PALs, ROMs, FPGAs, GALs, or CPLDs (see Glossary). A single RAM can replace or work in conjunction with the PLDs to produce nearly an unlimited number of different AMCs.

7. Quasi-Muller circuit designs of MAC module APSs are preferred. By designing MAC modules with C-elements as memory, use of input arbiters will usually not be necessary. Metastability may be a nonissue with properly designed Muller circuits. However, it is not clear if any C-element design can render it free of the metastable condition that may result when an input to a C-element is withdrawn before the "weak (keeper)" feedback inverter has fully responded to the last input change. See Section 3.7 and Chapter 11 for a more complete discussion of this subject.

9.3   C-ELEMENT DESIGN OF A 2 × 2 MAC MODULE

The state diagrams for the 2 × 2 SAM and TCM of a MAC module are given in Figure 9.3. Notice in the SAM that all state-to-state transitions are logically adjacent (Hamming distance of one). Also note that the inputs I_1 and I_0 indicate the state variable that changes for any given transition, and that the *transition enable inputs*, *To* and *Te*, are indicative of the parity of the *origin* state. Thus, TeI_1 indicates that the transition is from an *even parity* (EP) state and that the state variable of weight 2^1 will change during that transition. An even parity state is one that has an even number of 1's or no 1's. Obviously, an *odd parity* (OP) state is one that has an odd number of 1's. Reset is the inactive state of all inputs, meaning Reset $= \overline{I_1}\,\overline{I_0}$ in this case. It is also important to observe that the handshake interaction between the SAM and TCM forbid variable pairs *To, Te* and *So, Se* from becoming active simultaneously.

The NS and output K-maps for the two-input MAC module in Figure 9.3 are shown in Figure 9.4. Figure 9.4a and 9.4b shows the $Y \rightarrow SR$ conversion K-maps using the extended mapping algorithm in given in Section 2.1. (Note that the $Y \rightarrow S\overline{R}$ conversion algorithm cannot be used except for one-hot FSM design). The resulting SR K-maps and resulting logic are suitable for the C-element design of the MAC module (with the R input complemented), where the minimum logic is indicated by shaded loops. From the K-maps in Figure 9.4b and 9.4c, the optimum set of NS and output logic expressions for the SAM are given by Eqs. (9.1). Here, Reset $= \overline{I_1}\,\overline{I_0}$, OP $= y_1 \oplus y_0$, and EP $= \overline{y_1 \oplus y_0} = \overline{\text{OP}}$. Notice that static hazards are not possible in the NS logic expressions because only one *transition enable* (*Te* or *To*) can be active at any given time—two changing variables cannot produce a static hazard.

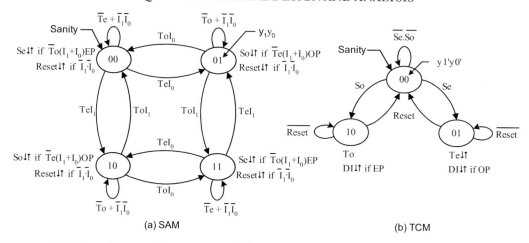

FIGURE 9.3: State diagrams for a two-input (2^2 state) MAC module. (a) The 2×2 state array machine (SAM) showing the branching conditions, holding conditions and outputs for even parity (EP) and odd parity (OP) states. (b) Timing control machine (TCM).

$$S_1 = \bar{y}_1\bar{y}_0\, TeI_1 + \bar{y}_1 y_0\, ToI_1 \quad R_1 = y_1\bar{y}_0\, ToI_1 + y_1 y_0\, TeI_1$$
$$S_0 = \bar{y}_1\bar{y}_0\, TeI_0 + y_1\bar{y}_0\, ToI_0 \quad R_0 = \bar{y}_1 y_0\, ToI_0 + y_1 y_0\, TeI_0$$
$$So = \bar{Te}\,(I_1 + I_0)(y_1 \oplus y_0) = \bar{Te} \cdot \overline{Reset} \cdot OP \qquad (9.1)$$
$$Se = \bar{To}\,(I_1 + I_0)(\overline{y_1 \oplus y_0}) = \bar{To} \cdot \overline{Reset} \cdot EP$$

The optimized NS and output logic expressions for the TCM are obtained directly from the state diagram in Figure 9.3b and from Figure 9.4d and 9.4e, respectively, and are given in Eqs. (9.2). Direct reading of the TCM state diagram, to give its NS logic, is possible because So and Se cannot be active at the same time and $R_1 = R_0 = Reset$ are required to return the TCM to the unresolved state, 00. Thus, the subscripted variables S_1, R_1 and S_0, R_0 are the inputs to the TCM C-elements 1 and 0, respectively. The prime symbol on TCM state variables, y'_1 and y'_0, is used to distinguish them from the SAM state variables, y_1 and y_0.

$$\left\{ \begin{array}{ll} S_1 = So & S_0 = Se \\ R_1 = Reset & R_0 = Reset \\ \multicolumn{2}{c}{To = y'_1} \\ \multicolumn{2}{c}{Te = y'_0} \\ \multicolumn{2}{c}{DI = y'_1 EP + y'_0 OP} \\ \multicolumn{2}{c}{= ToEP + TeOP} \end{array} \right\} \qquad (9.2)$$

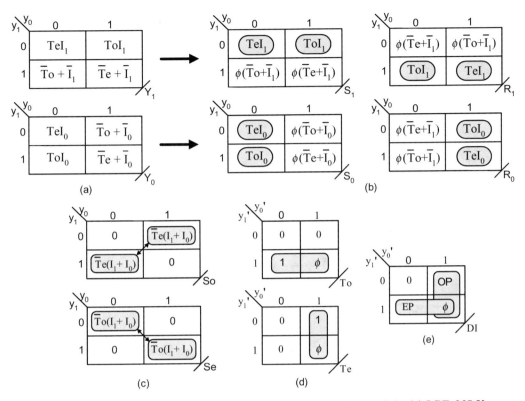

FIGURE 9.4: NS and output K-maps for the design of a 2 × 2 MAC module. (a) LPD NS K-maps as plotted from Figure 9.3a. (b) The Y → SR K-map conversion algorithm in Section 2.1 for the C-element design showing minimum cover. (c) Output K-maps showing XOR-type configurations for a minimum representation. (d and e) Output K-maps for the TCM.

The optimum logic circuit for the two-input (four-state) MAC module is shown in Figure 9.5, where C-elements are used as the memory elements. When C-elements are used as the memory, metastability becomes a nonissue except as discussed in Section 3.7. The fact that C-elements operate outside of the fundamental mode is important to the reliability of a MAC module and to any APS in which it is used. The 2 × 2 MAC module logic in Figure 9.5 has but one exclusive OR (XOR) gate to generate the OP signal. This XOR gate can be optimized for speed by using a modern CMOS version consisting of only six high-speed transistors (see Tinder's books in Endnotes). Keep in mind that the internal handshake mechanism depends on the parity parameters EP and OP, where for an n-input MAC module OP $= y_{n-1} \oplus \cdots \oplus y_2 \oplus$

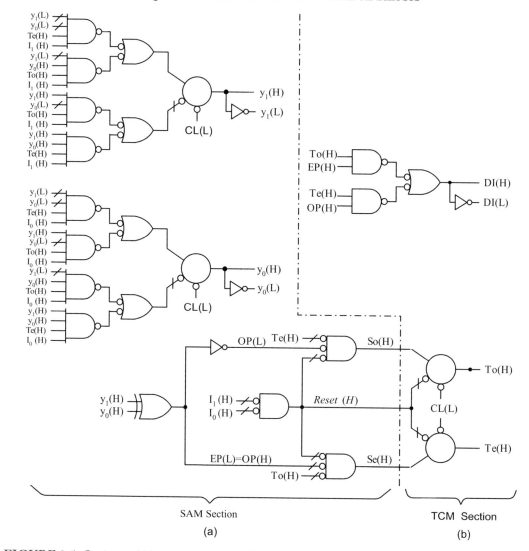

FIGURE 9.5: Optimum NS and output logic diagrams for the C-element design of a two-input (four-state) MAC module. (a) SAM section implemented from Eqs. (9.1). (b) TCM section implemented directly from the state diagram in Figure 9.3b and Eqs. (9.2).

$y_1 \oplus y_0$. Thus, the larger the state capacity of the MAC module, the slower will be its state-to-state transition response time. Remember that a string of XORed functions requires an XOR tree. As will be demonstrated in Section 9.4, two-input MAC modules can be cascaded (according to Table 9.1) without compromising speed or reliability, a very important design consideration. MAC modules having inputs grater than 2 are unnecessary due to the cascading capability of the two-input module.

9.3.1   Stepwise Operation of the MAC Module

Before applying the MAC module to the design of asynchronous FSMs, let us step through the operation of the two-input MAC by using its state diagrams in Figure 9.3. The reader should follow this closely. Begin with the initialization of both SAM and TCM machines into their 00 states. In the 00 state, the SAM is in an even parity state, so EP is active, $EP(H) = 1(H)$. Also, because neither transition enable is activated then Te is inactive and $\overline{Te}$ is active. However, output Se cannot yet be activated because neither input (I_1 or I_0) has been activated. One or the other of these inputs, but not both, will be activated according to the program table valid for the specific FSM to be implemented with the MAC module. Now let us say I_1 becomes active. This allows Se to be issued by the SAM which, in turn, causes the TCM to transit $00 \rightarrow 01$ issuing Te unconditionally in state 01. This permits the SAM to transit $00 \rightarrow 10$ under conditions TeI_1, which is now active. Remember that a transition enable refers to the state from which the transition originates. Upon

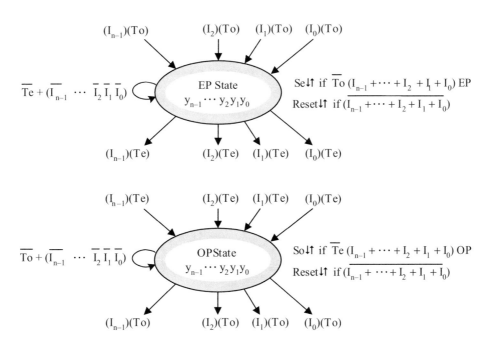

FIGURE 9.6: Generalized transition conditions and outputs for the EP and OP states of a 2^n state SAM with n-way out-branching.

arrival of the SAM in state 10, where OP becomes active, the TCM output DI is activated causing Reset to be issued by the SAM, which in turn causes the TCM to transit back $01 \rightarrow 00$ deactivating Te. Now in state 10, the SAM awaits for the activation of a single input (I_1 or I_0) before output So can be issued. When one of these two inputs becomes active, the SAM will transit to an EP state, depending on the particular input that becomes active, and the process begins all over again—an orderly, well-defined handshake process.

It is important to remember that each state of the SAM is *logically adjacent* to all states to which it can transit. This fact eliminates any chance that a critical race or output race glitch can occur. Moreover, it is the programming of the MAC module that permits only one input to be active at any given time, thereby eliminating static, function and dynamic hazards from forming in the NS logic of the SAM. The generalized transition conditions and outputs for the EP and OP states for a 2^n state SAM with n-way branching is shown in Figure 9.6.

9.4   CASCADING THE MAC MODULES

Table 9.1 provides the various cascading possibilities for MAC modules. Shown in Figure 9.7 is an example of two 2×2 MAC modules that are cascaded (combined) in parallel to produce a 4×4 MAC module. The 4×4 MAC module has a 16-state capability with up to four-way branching and possesses the same speed and reliability as each of its component 2×2 MAC modules. Three 2×2 MAC modules can be cascaded in parallel to produce a MAC module with a 64-state capacity with up to six-way branching capability all without compromising speed and reliability. Alternative cascading possibilities are given in Table 9.1 as discussed previously.

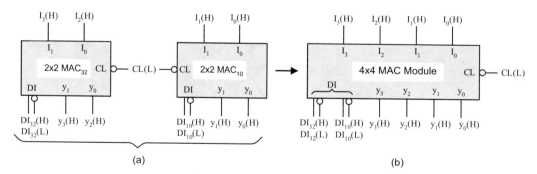

(a)   (b)

FIGURE 9.7: (a) Two 2×2 MAC modules each with four-state capability and up to two-way branching. (b) The two 2×2 MAC modules in (a) combined to produce a 4×4 MAC module with 16-state capability and up to four-way branching with no loss of speed or reliability.

9.5   PROGRAMMING THE MAC MODULE—FOUR EXAMPLES

We will now give examples of four FSMs that are programmed for implementation by a single 4×4 MAC module consisting of two cascaded 2×2 MAC modules as in Figure 9.7. Each FSM will have a state diagram, a program table derived from the state diagram, and the minimum logic as obtained from K-maps by using standard mapping procedures. Some of the state diagram representations will use buffer states to maintain logically adjacent transitions as required by MAC modules. Others will require the use of the deactivate inputs (DI) to maintain orderly control of the transitions. Normally, involvement of the DI feature will not be necessary for FSMs whose state-to-state transitions are controlled by external inputs. It is the presence of cycle states with no external input control that will require the use of the DI feature. Buffer states are not cycle states and do not need the use of the DI feature because the transition to a buffer state is controlled by an external input.

Shown in Figure 9.8 is the state diagram (a), the PS and NS instructions program table (b), and conventional third-order K-maps (c), all for the MAC-0 module design of a 3-bit Gray code counter. Notice that none of the state-to-state transitions are controlled by external inputs, which makes it necessary to use the DI feature for control purposes. Without the use of the DI feature, this

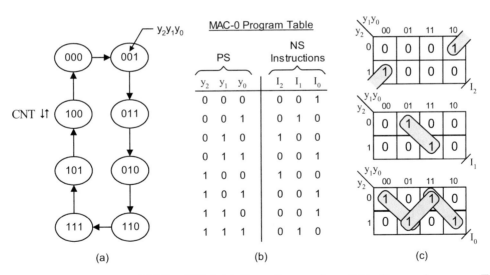

FIGURE 9.8: Design of the MAC-0 FSM. (a) State diagram for a 3-bit Gray code counter FSM. (b) Program table of the 3-bit Gray code counter in (a) as required for implementation by the 4×4 MAC module in Figure 9.7. (c) Optimum XOR logic from the table in (b) suitable for MAC-0 module implementation.

FSM would go out of control much like a multistate oscillator. Recall that as a synchronous FSM, this counter would be controlled by the action of a system clock input to three flip-flops.

The optimum XOR-type logic cover is indicated by shaded loops in Figure 9.8c. The optimum results are easily read directly from the K-maps and are given by Eqs. (9.3). In these expressions, the equivalence symbol is given by $\overline{\oplus}$ (complement of the XOR symbol). For a review of XOR algebra and graphics, see Appendix A.2 and Tinder's text in Endnotes.

$$I_2 = \bar{y}_0(y_2 \oplus y_1) \quad I_1 = y_0(\overline{y_2 \oplus y_1}) \quad I_0 = \overline{y_2 \oplus y_1 \oplus}$$
$$\text{CNT} = y_2\bar{y}_1\bar{y}_0 \tag{9.3}$$

The optimum logic circuit for the 3-bit Gray code counter is shown in Figure 9.9 in agreement with Eqs. (9.3). The outputs are enabled with NOR gates such that an $EN(L) = 1(L)$ enables the output, whereas an $EN(L) = 0(L)$ disables them. Control of output activation in this manner is essential to the time-shared multiplexing of different FSMs by using a single MAC module. Note that the gate/input tally for the XOR/NAND/NOR circuit in Figure 9.9 is 4/8 (excluding the enabling and output logic) as compared to 11/32 for two-level sum-of-product (SOP) logic, again excluding the enabling and output logic. Obviously, there is a considerable savings in hardware by using XOR-type logic. Note that the deactivate inputs, DI_{32} and DI_{10} are necessary for MAC module control of the cycle states.

As a second example, consider the FSM in Figure 9.10 that will detect the direction of rotation [counterclockwise (CCW) or clockwise (CW)] of a cylindrical shaft under variable rotational speed. This is accomplished by photocell sensing of light beams reflected off the shaft's end surface, half of which has been made a reflecting and half nonreflecting. Thus, the rotational speed of the shaft is limited to the response time of the photocells. Shown in Figure 9.10(a) is the MAC-1

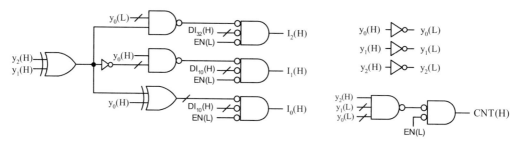

FIGURE 9.9: MAC-0 implementation of Eqs. (9.3) with XOR/NAND/NOR logic with active low enable $EN(L)$ and DI inputs and an active high output, $CNT(H)$, for use with the 4 × 4 MAC module in Figure 9.7b.

FSM state diagram and inputs as required for detecting the rotational direction of the shaft. Figure 9.10b shows the program table read directly from the state diagram in Figure 9.10a suitable for use with either a 2 × 2 MAC module or the 4 × 4 MAC module given in Figure 9.7b. Note that the input requirements are those needed for the FSM to transit to logically adjacent states with a change of only one state variable. For example, in present state 00, an NS instruction of $I_1 = A\overline{B}$ is needed to effect the transition $00 \rightarrow 10$ where only state variable y_1 changes. Or, in PS 10, a NS instruction $I_0 = \overline{A}\,\overline{B}$ causes a transition $10 \rightarrow 11$ that takes place with a y_0 state variable change. Continuing in this manner, the MAC-1 transitions occur strictly controlled by the MAC module, free of all timing defects including possible E-hazards.

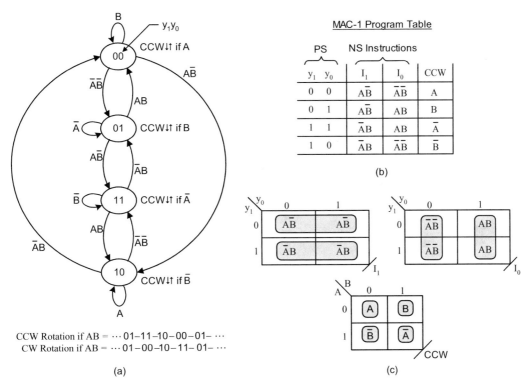

FIGURE 9.10: Design of the MAC-1 FSM required to detect the direction of rotation (CCW or CW) of a cylindrical shaft. (a) State diagram and input requirements for the rotation detector having two inputs and one output. (b) Program table as required to implement the MAC-1 FSM by using the 4 × 4 MAC module in Figure 9.7. (c) EV K-maps showing the optimum two-level input and output logic suitable for MAC-1 implementation.

The entered variable (EV) K-maps for the NS instructions, I_1 and I_0, and the single output CCW are given in Figure 9.10c. The optimum two-level logic expressions for the NS instructions and output are indicated by shaded loops and are given by Eqs. (9.4) and (9.5), respectively. Also given in Eqs. (9.4) are the XOR-type expressions for the two NS instruction parameters, which we will use in preference to the two-level logic. The XOR-type expressions in Eqs. (9.4) can be looped out directly in maxterm code from their respective K-maps. (See Tinder's text in Endnotes for K-map minimization of XOR-type functions.) Note that there are two internally initiated static-1 hazards in the CCW expression and that they are covered by the two p-terms shown. For a review of static hazards in the NS and output logic expressions, refer to Section 3.3.

$$I_1 = \bar{y}_1 A\bar{B} + y_1\bar{A}B = \left(\bar{y}_1 A + y_1\bar{A}\right)\left(\bar{y}_1\bar{B} + y_1 B\right) = \left(y_1 \oplus A\right)\left(y_1 \overline{\oplus} B\right)$$
$$I_0 = \bar{y}_0\bar{A}\bar{B} + y_0 AB = \left(\bar{y}_0\bar{A} + y_0 A\right)\left(\bar{y}_0\bar{B} + y_0 B\right) = \left(y_0 \overline{\oplus} A\right)\left(y_0 \overline{\oplus} B\right)$$

$$(9.4)$$

$$CCW = \bar{y}_1\bar{y}_0 A + \bar{y}_1 y_0 B + y_1\bar{y}_0\bar{B} + y_1 y_0\bar{A} + \underbrace{\bar{y}_1 AB + y_0\bar{A}B}_{\text{Hazard Cover}}$$

$$(9.5)$$

Implementation of the logic expressions in Eqs. (9.4) and (9.5) is given in Figure 9.11, where NOR gates are used for enable NS instruction purposes. The NS instruction inputs, I_1 and I_0, are

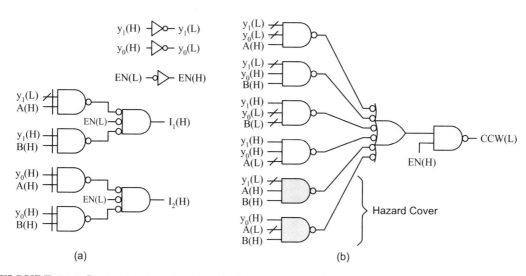

FIGURE 9.11: Logic circuits using the wireless connection feature for the MAC-1 FSM consistent with Eqs. (9.4) and (9.5), and suitable to be implemented by the 4 × 4 MAC module shown in Figure 9.7. (a) NS instruction logic circuit made up of XOR/NOR logic with active low enable EN(*L*). (b) NAND output logic circuit with EN(*H*), active low output CCW(*L*) and showing hazard cover indicated by shaded gates.

implemented by using exclusively XOR/NOR logic. For a review of mixed-logic XOR and EQV gate symbology, see Appendix A.1. Hazard cover in the output CCW is shown as shaded NAND gates.

The third example is that for the MAC-2 FSM featuring a five-state FSM with an additional two states used as buffer states to permit logically adjacent transitions as required by MAC module implementation. This FSM has two inputs, S and T, controlling the state-to-state transitions and

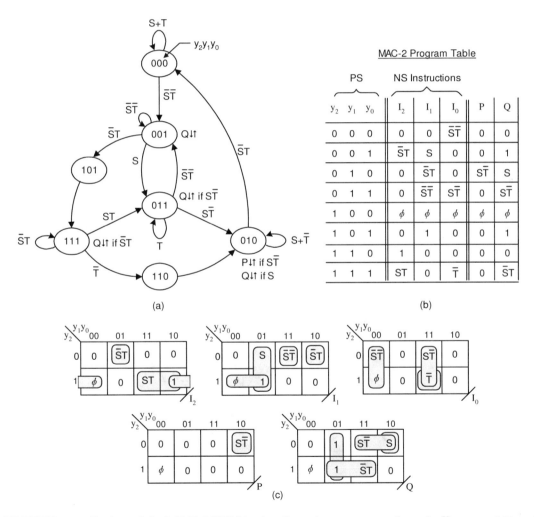

FIGURE 9.12: Design of the MAC-2 FSM having five primary states and two buffer states, 101 and 110, as required to prodice logically adjacent transitions. (a) State diagram. (b) Program table obtained directly from the state diagram in (a). (c) NS instruction and output K-maps obtained from the program table in (b) and showing optimum cover.

two outputs, P and Q, as shown in the state diagram of Figure 9.12a. The program table for the MAC-2 FSM is obtained directly from the state diagram and is presented in Figure 9.12b. From the program table, the NS instruction and output functions are plotted in EV K-maps shown in Figure 9.12c, where optimum cover is indicated by shaded loops. The NS and output expressions for the MAC-2 FSM are extracted from the optimum cover in Figure 9.12c and are presented by

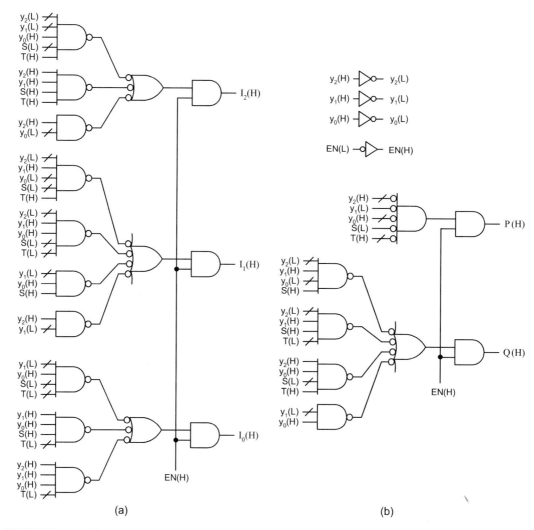

(a)   (b)

FIGURE 9.13: Logic circuits using the wireless connection feature for the MAC-2 FSM consistent with Eqs. (9.6) and suitable to be implemented with the 4×4 MAC module given in Figure 9.7. (a) NS instruction logic with NAND/AND logic. (b) Output circuits with NOR-NAND/AND logic.

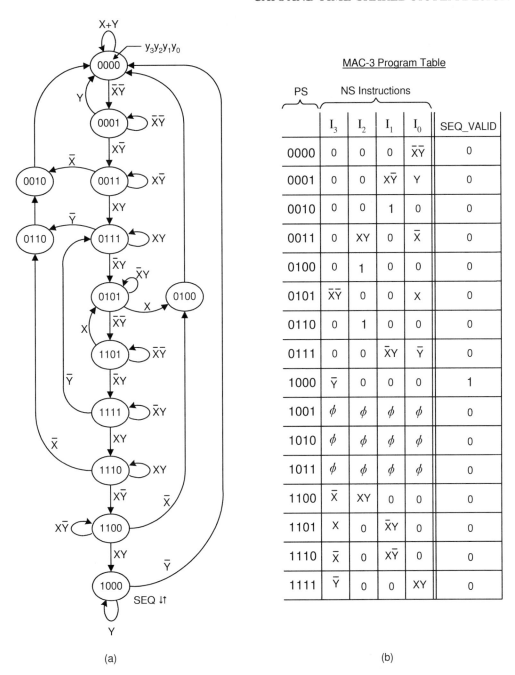

MAC-3 Program Table

PS	NS Instructions				
	I_3	I_2	I_1	I_0	SEQ_VALID
0000	0	0	0	$\overline{X}\overline{Y}$	0
0001	0	0	$X\overline{Y}$	Y	0
0010	0	0	1	0	0
0011	0	XY	0	$\overline{X}$	0
0100	0	1	0	0	0
0101	$\overline{X}\overline{Y}$	0	0	X	0
0110	0	1	0	0	0
0111	0	0	$\overline{X}Y$	$\overline{Y}$	0
1000	$\overline{Y}$	0	0	0	1
1001	ϕ	ϕ	ϕ	ϕ	0
1010	ϕ	ϕ	ϕ	ϕ	0
1011	ϕ	ϕ	ϕ	ϕ	0
1100	$\overline{X}$	XY	0	0	0
1101	X	0	$\overline{X}Y$	0	0
1110	$\overline{X}$	0	$X\overline{Y}$	0	0
1111	$\overline{Y}$	0	0	XY	0

(a)   (b)

FIGURE 9.14: Design of the MAC-3 FSM, a 10-state sequence recognizer having two inputs controlling logically adjacent transitions and one output. (a) State diagram involving three direct buffer states and two indirect buffer states. (b) Program table for the PS and NS instruction parameters, and the single output.

Eqs. (9.6). Note that static hazards are not possible in the NS instruction functions and none are indicated in the output expressions. Implementation of Eqs. (9.6) is given in Figure 9.13, where again the wireless connection feature is used to simplify the schematic capture.

$$I_2 = \bar{y}_2\bar{y}_1 y_0 \bar{S}T + y_2 y_1 ST + y_2\bar{y}_0$$

$$I_1 = \bar{y}_2 y_1\bar{y}_0 \bar{S}T + \bar{y}_2 y_1 y_0 \bar{S}\bar{T} + \bar{y}_1 y_0 S + y_2\bar{y}_1$$

$$I_0 = \bar{y}_1\bar{y}_0 \bar{S}\bar{T} + y_1 y_0 S\bar{T} + y_2 y_1 y_0\bar{T}$$

$$P = \bar{y}_2 y_1\bar{y}_0 S\bar{T}$$

$$Q = \bar{y}_2 y_1\bar{y}_0 S + \bar{y}_2 y_1 S\bar{T} + y_2 y_0 \bar{S}\bar{T} + \bar{y}_1 y_0$$

(9.6)

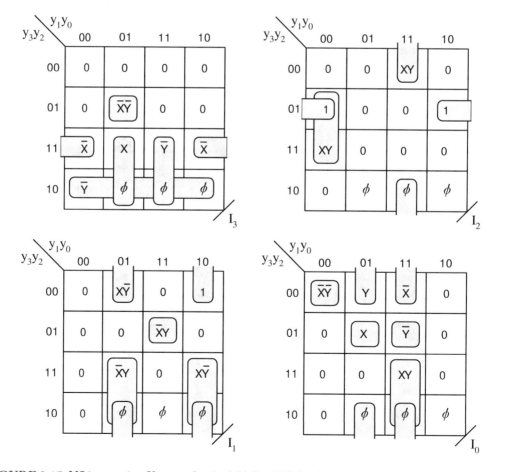

FIGURE 9.15: NS instruction K-maps for the MAC-3 FSM in Figure 9.14a plotted from the program table in Figure 9.14b showing optimum cover indicated by shaded loops.

As our final example, consider the state diagram and program table for the MAC-3 FSM, a 10-state sequence recognizer suitable for implementation by the 4 × 4 MAC module shown in Figure 9.7a. The state diagram in Figure 9.14a has two inputs controlling the state transitions and

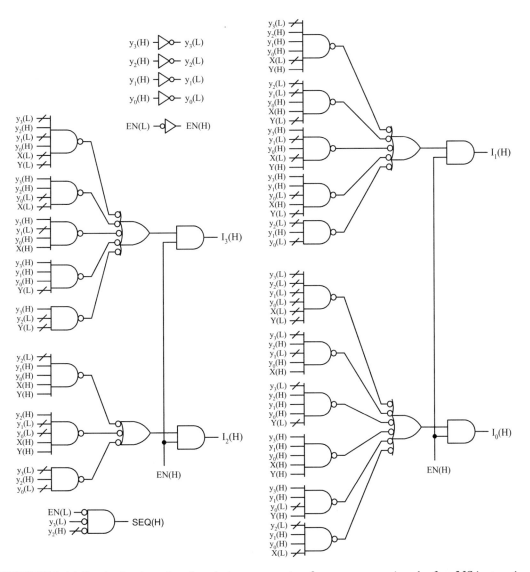

FIGURE 9.16: Logic circuits using the wireless connection feature representing the four NS instruction inputs and the output function SEQ with enable for the MAC FSM-3 in agreement with the K-map logic cover given in Figure 9.15 and Eqs. (9.7), and suitable for implementation by using the the 4 × 4 MAC module in Figure 9.7.

one output issued in state 1000 at the end of a successful sequence. Any violation of the required sequence must return the FSM to the initiation state, 0000, whereupon the FSM automatically transits to the first state in the sequence, 0001. Note that three direct buffer states (0010, 0100, and 0110) and two indirect buffer states (0101 and 0111) are required to produce logically adjacent (reverse parity) transitions given the state assignment shown. By retaining the 0000 initiation state, alternative state code assignment schemes are acceptable as long as reverse parity transitions are ensured and the FSM is initialized into the 0000 state.

Shown in Figure 9.15 are the EV K-maps for the NS instruction parameters with the optimum cover indicated by shaded loops. Note that there are no shared prime implicants typical for MAC module designs. The hardware requirements are significant and perhaps best implemented by using a PLD such as a PLA. We will use the optimum two-level SOP logic indicated by shaded loops in the K-maps. This logic extracted from the K-maps is presented in Eqs. (9.7) indicating fan-in requirements from three to six. The output SEQ (sequence valid) in state 1000 is given simply by $y_3\bar{y}_2$ making use of the three don't care states. Note that static hazards are not possible in the NS instruction logic because the handshake interaction between the SAM and TCM does not permit them to exist.

$$I_3 = \bar{y}_3 y_2 \bar{y}_1 y_0 \bar{X}\bar{Y} + y_3 y_2 \bar{y}_0 \bar{X} + y_3 \bar{y}_1 y_0 X + y_3 y_1 y_0 \bar{Y} + y_3 \bar{y}_2 \bar{Y}$$
$$I_2 = \bar{y}_2 y_1 y_0 XY + y_2 \bar{y}_1 \bar{y}_0 XY + \bar{y}_3 y_2 \bar{y}_0$$
$$I_1 = \bar{y}_3 y_2 y_1 y_0 \bar{X}Y + \bar{y}_2 \bar{y}_1 y_0 X\bar{Y} + y_3 \bar{y}_1 y_0 \bar{X}Y + y_3 y_1 \bar{y}_0 X\bar{Y} + \bar{y}_2 y_1 \bar{y}_0 \qquad (9.7)$$
$$I_0 = \bar{y}_3 \bar{y}_2 \bar{y}_1 \bar{y}_0 \bar{X}\bar{Y} + \bar{y}_3 y_2 \bar{y}_1 y_0 X + \bar{y}_3 y_2 y_1 y_0 \bar{Y} + y_3 y_1 y_0 XY + \bar{y}_2 \bar{y}_1 y_0 Y + \bar{y}_2 y_1 y_0 \bar{X}$$
$$\text{SEQ} = y_3 \bar{y}_2$$

9.6 TIME-SHARED FSM OPERATION BY USING CASCADED MAC MODULES

In Figure 9.17, we illustrate how a single 4×4 (16-state) MAC module, following the architecture of Figure 9.2, can be used to drive four independent, radically different FSMs on a time-shared basis all defect-free. These state machines vary in complexity from an eight-state Gray code generator to a 10-state sequence recognizer. The state diagrams, program tables, and logic circuits for the four FSMs are given in Section 9.5, each illustrating something different. They are named MAC-0, MAC-1, MAC-2, and MAC-3. For simplicity, these logic circuits, including Figure 9.17, are constructed by using the wireless connection feature that is emphasized in this book. Optimum logic for each FSM is obtained from EV K-maps following standard mapping procedures. The interfacing logic in Figure 9.17 is necessary to generate I_2, I_1, and I_0 but not I_3 because only MAC FSM-3 requires I_3, hence $3I_3 = I_3$. The select inputs to the 2-4 decoder, D_0, D_1, select which FSM of the four is to be activated at any given time and operated by the 4×4 MAC module.

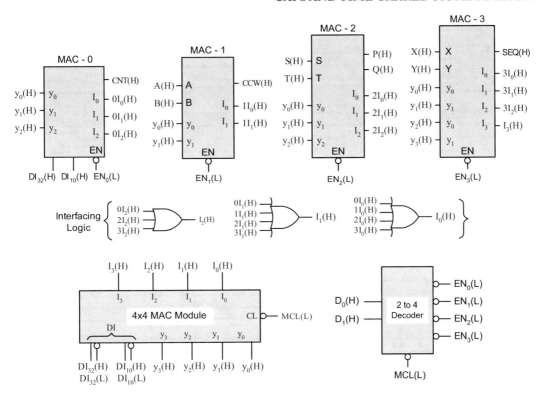

FIGURE 9.17: Block diagram architecture, consistent with Figure 9.2, used to illustrate the time-shared operation of four different FSMs, MAC-0, MAC-1, MAC-2, and MAC-3, by using the 4 × 4 MAC module in Figure 9.7b with master CLEAR (MCL).

Figure 9.18 illustrates the simulation of the four FSM architecture in Figure 9.17 showing the time-shared results for FSMs MAC-0, MAC-1, MAC-3, and MAC-2 presented in that order. It is important to note that only for MAC FSM-0 is it necessary to use deactivate inputs, DI_{32} and DI_{10}, for MAC module control because all state-to-state transitions are *cycles* not controlled by external inputs. The remaining three FSMs have state-to-state transitions all controlled by external inputs. MAC FSM-2 has two *buffer states* (101 and 110) and MAC FSM-3 has three buffer states, all of which are controlled by external inputs. Thus, one cycle state is acceptable and need not be controlled by DIs to the FSM instruction logic. Note that in MAC FSM-3 the primary state 0111 also serves as a buffer state controlled by the external input $\bar{Y}$. Obviously, MAC FSM-1 has neither buffer nor cycle states. It would not be wrong to use the DI feature for all FSMs, but that would be unnecessary and would lead to additional inputs thereby increasing throughput delay. Note that the inputs to the four FSMs are continuously present during the entire time-shared operation, but

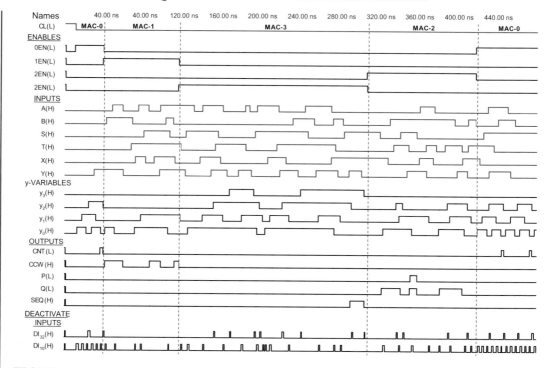

FIGURE 9.18: Time-shared simulation of four radically different asynchronous FSMs featured in Figure 9.17 all driven by the 4 × 4 MAC module shown in Figure 9.7b as enabled by a 2-4 decoder.

the corresponding outputs are active only when the given FSM is enabled. Remember that all y variables can be deactivated anytime during the time-shared operation of the four FSMs by using the master CLEAR. It would be more complicated to activate only the y variables to the FSM selectively enabled.

The DI(H) outputs from the 4 × 4 MAC module in Figure 9.17 mark each state-to-state transition in a given time-shared operation and are shown in Figure 9.18. As pointed our earlier, such DI signals can be used to control system peripherals or other operations as, for example, transition count and transition time analysis.

What we have just demonstrated is the use of a MAC module (a cascaded MAC module system in this case) to switch instantly from one asynchronous FSM to another radically different one on a time-shared basis—all free of timing defects. This is quite remarkable because timing defects owned exclusively by asynchronous FSMs are normally considered a major setback to their design and operation. Combinational hazards and sequential hazards (essential hazards), which are timing defects with the potential to cause malfunction, are not possible in a MAC module FSM design. What this means is that a designer can design and operate any asynchronous FSM and be rest assured that it will operate correctly as designed. That is the upside of the MAC approach. The downside is that each state-to-state transition must be logically adjacent, a Hamming distance of one. This, in turn, may require the use of additional state variables and the insertion of buffer and cycle states. But the logically adjacent transition requirement eliminates critical races and output race glitches, removing those issues from consideration.

Clearly, use of the MAC module to operate a number of controller FSMs on a time-shared basis offers an attractive opportunity to design complex asynchronous systems all free of timing defects. By logically combining the appropriate DI and y variable signals from the MAC module, it is possible to activate specific data path devices when required for each of the controllers.

CHAPTER 10

Asynchronous One-Hot Programmable Sequencer Systems

Before continuing, the reader should review Chapter 5, which provides the necessary background to understand one-hot asynchronous sequencers. In particular, Eqs. (5.1) and the characteristics of the one-hot method outlined in Section 5.2 should be thoroughly understood before moving on to the more complex subject matter in this chapter.

The one-hot programmable sequencer (A-OPS) enjoys some attractive advantages over the microprogrammable asynchronous controller (MAC) module approach discussed in the previous chapter. Because of the one-hot coding (one "1" for each state), a timing control machine is not needed—no parity detection or deactivation of inputs is required. Furthermore, programming of a one-hot sequencer is exceedingly simple because it is only necessary to provide the sequencer with the branching condition for each one-hot state-to-state transition as read from a state table or state diagram of the finite state machine (FSM) to be designed. Similar to the MAC module, the one-hot approach requires a single-state array sequencer machine that can support any number of FSMs on a time-shared basis, but only if the FSMs do not exceed the state number limitation of the sequencer. Unlike the MAC module sequencer, a one-hot sequencer can support any state-to-state transition in an FSM provided it is *void of cycle conditions*, including, in particular, endless cycles. Recall that the deactivate inputs DI feature of the MAC module permits cycles to exist and be controlled.

10.1 GENERAL ARCHITECTURE

The generalized architecture for one-hot asynchronous programmable sequencers is shown in Figure 10.1 and is predictably similar to that for the MAC module given in Figure 9.2. The only difference is that the 2^n state MAC module in Figure 9.2 is replaced by an n-state one-hot sequencer with $n^2 2^k$ input interfacing logic usually implemented with OR gates. However, there is one important difference in the manner in which these sequencers are used. One-hot asynchronous sequencers cannot be cascaded like MAC modules in Figure 9.7. Thus, only a single one-hot sequence is permitted to operate a given number of FSMs on a time-shared basis, where each FSM is limited to the n states of the sequencer.

An inspection of Figure 10.1 indicates that an n-state one-hot sequencer requires specification of n^2 inputs, one for each branching condition in an $n \times n$ state array. As indicated in Figure 10.2, each jth state of a completely specified n-state one-hot sequencer requires n input branching paths, including the required holding condition. This means there exists n-way branching capability to and from each state. Thus, for n states, n^2 branching conditions must be specified in a one-hot sequencer that contains all possible branching paths. For example, a 10-state one-hot sequencer requires that 100 branching conditions be specified for a given FSM design, although many of these branching conditions are set to logic 0 if their corresponding branching paths do not exist for the implementation of a given FSM. Clearly, this is a hardware-intensive design. To avoid costly fan-in

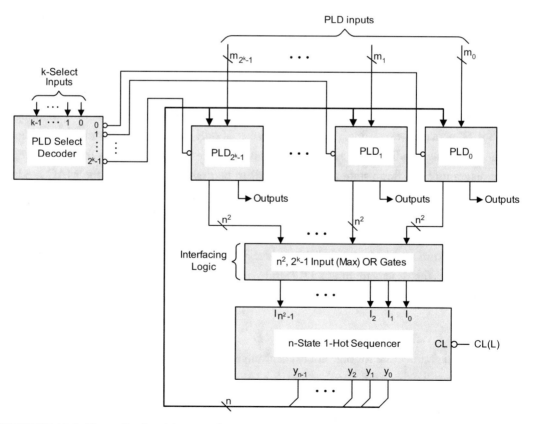

FIGURE 10.1: Generalized architecture for programming an asynchronous n-state one-hot sequencer with n^2 inputs with feedback y variables to 2^k - 1 PLDs.

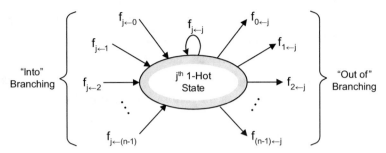

FIGURE 10.2: Generalized transition conditions for the jth state of an n-state one-hot sequencer showing n-way branching to and from the jth state.

delays due to multiple input ORing operations, use can be made of the complementary metal-oxide semiconductor (CMOS) NOR gate in Figure 8.2 with active low inputs and with an inverter on its output. Such a NOR gate is capable of accommodating up to eight inputs without significant throughput delays. For large n-state one-hot sequencers, use of modern programmable logic devices (PLDs) is recommended, preferably programmable logic arrays (PLAs).

An FSM designed via the one-hot method has but a single "1" assigned to each unique state of the FSM as mentioned earlier. This makes possible the use of alphabetic or numeric state identifiers such as a, b, c, ... or 0, 1, 2, ... Thus, use of state code assignments is not necessary or even desirable. The general next state (NS) and output equations for a single one-hot FSM of m states are given by Eqs. (5.1). For an n-state A-OPS, there must be n^2 branching conditions such that the NS equations are represented generally as

$$Y_j = \underbrace{\sum_{j=0}^{n-1}\sum_{i=0}^{n-1} f_{ij} y_i}_{\text{"Into" Terms}} + \underbrace{y_j \cdot \overline{\sum_{\substack{k=0 \\ k \neq j}}^{n-1} y_k}}_{\text{"Out of" Terms}} \qquad (10.1)$$

with no output Z_l expressions required for an A-OPS. In quasi-tensor subscript notation, by using the Einstein summation convention, Eq. (10.1) can be written more succinctly as

$$Y_j = f_{ij} y_i + y_j \overline{F_j} \qquad i,j = 0,1,2,\ldots, n-1 \qquad (10.2)$$

Here, F_j represents the Boolean sum of all y variables to which the jth state transits and $\overline{F_j}$ is the complement of that sum. Also, it should be noted that $f_{ij} \neq f_{ji}$, meaning that $f_{j \leftarrow i} \neq f_{i \leftarrow j}$, that is, the branching condition matrix is asymmetric (not symmetric). The "out of" term for each NS function is a p-term consisting of the uncomplemented state variable for that function ANDed with the complement of the remaining state variables. Note that an *initialization* term $\bar{y}_0 \bar{y}_1 \bar{y}_2 \bar{y}_3 \ldots \bar{y}_{n-1}$ must be combined with a specific Y_j in Eq. (10.1) for use with the *one-hot-plus-zero approach* to initialization. This is accomplished by using the factoring and absorptive laws (see Appendix A.2) resulting in a reduced p-term consisting of the complement of all state variables exclusive of that for the initialization state. Once the one-hot-plus-zero implementation is complete, the sanity circuit can be used to drive the FSM (to be designed) into the initial one-hot state via the all-zero state.

10.2 DESIGN OF ONE-HOT SEQUENCERS

To illustrate, we design a fully specified four-state one-hot sequencer having the state diagram shown in Figure 10.3. Here, initialization into state 0 is accomplished by the one-hot-plus-zero method previously described in Section 5.2 and illustrated in Section 5.3. Note that each state requires specification of four branching conditions each represented in the form f_{ij}.

Application of Eq. (10.2) to a four-state A-OPS results in the matrix Eq. (10.3) composed of the "into" terms and the "out of" terms as indicated. Notice that the first of the "out of" terms on the right side of Eq. (10.3) results from the factoring and absorptive laws, given in Appendix A.2, applied to the sum of the "out of" and "1-Hot + Zero" terms as required for initialization into the "0" state. Thus, $y_0 \bar{y}_1 \bar{y}_2 \bar{y}_3 + \bar{y}_0 \bar{y}_1 \bar{y}_2 \bar{y}_3 = \bar{y}_1 \bar{y}_2 \bar{y}_3$.

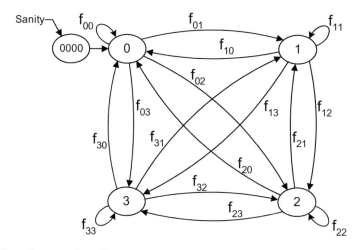

FIGURE 10.3: State diagram for a four-state 1-hot sequencer that will initialize into the state 0 via the one-hot-plus-zero method.

$$
\begin{bmatrix} Y_0 \\ Y_1 \\ Y_2 \\ Y_3 \end{bmatrix} = \underbrace{\begin{bmatrix} f_{00}\, f_{10}\, f_{20}\, f_{30} \\ f_{01}\, f_{11}\, f_{21}\, f_{31} \\ f_{02}\, f_{12}\, f_{22}\, f_{32} \\ f_{03}\, f_{13}\, f_{23}\, f_{33} \end{bmatrix}}_{\text{“Into” Terms}} \begin{bmatrix} y_0 \\ y_1 \\ y_2 \\ y_3 \end{bmatrix} + \underbrace{\begin{bmatrix} \bar{y}_1 \bar{y}_2 \bar{y}_3 \\ y_1 \bar{y}_0 \bar{y}_2 \bar{y}_3 \\ y_2 \bar{y}_0 \bar{y}_1 \bar{y}_3 \\ y_3 \bar{y}_0 \bar{y}_1 \bar{y}_2 \end{bmatrix}}_{\text{“Out of” Terms}}
\tag{10.3}
$$

Expansion of the matrices in Eq. (10.3) results in the equations for the lumped path delay design of the asynchronous one-hot sequencer. However, we will use Muller C-elements (Figure 1.12 or 1.13) in the design of A-OPS machines because C-elements operate outside of the fundamental mode. To do this requires that we apply the simple $Y \rightarrow S\bar{R}$ algorithm given by Eqs. (5.2) and is restated below for the convenience of the reader:

$$
\left\{ \begin{array}{l} S_j = \sum (\text{sum}) \text{ of all non-}y_j \text{ p-terms in } Y_j \\ \bar{R}_j = \sum \text{of all } y_j \text{ p-term coefficients in } Y_j \end{array} \right\} \quad Y \rightarrow S\bar{R} \text{ Algorithm}
\tag{5.2}
$$

Here, it is seen that the $\bar{R}_j$ function includes the holding condition f_{jj} and the ANDed complements of all the present state variables except that for y_j. The exception is $\bar{R}_0$, which is always f_{00} with the application of the one-hot-plus-zero initialization method. By using this conversion algorithm, we construct the S_j and $\bar{R}_j$ equations directly from Eq. (10.3) to produce the results given by Eqs. (10.4).

$$
\begin{aligned}
S_0 &= f_{10}y_1 + f_{20}y_2 + f_{30}y_3 + \bar{y}_1 \bar{y}_2 \bar{y}_3 \\
\bar{R}_0 &= f_{00} \\
S_1 &= f_{01}y_0 + f_{21}y_2 + f_{31}y_3 \\
\bar{R}_1 &= f_{11} + \bar{y}_0 \bar{y}_2 \bar{y}_3 = f_{11} + \overline{(y_0 + y_2 + y_3)} \\
S_2 &= f_{02}y_0 + f_{12}y_1 + f_{32}y_3 \\
\bar{R}_2 &= f_{22} + \bar{y}_0 \bar{y}_1 \bar{y}_3 = f_{22} + \overline{(y_0 + y_1 + y_3)} \\
S_3 &= f_{03}y_0 + f_{13}y_1 + f_{23}y_2 \\
\bar{R}_3 &= f_{33} + \bar{y}_0 \bar{y}_1 \bar{y}_2 = f_{33} + \overline{(y_0 + y_1 + y_2)}
\end{aligned}
\tag{10.4}
$$

We can apply the same procedure to the design of any n-state A-OPS. As a second example, consider the six-state asynchronous one-hot sequencer fully specified by the state diagram in Figure 10.4. Here, the 36 branching functions f_{ij} are shown together with the Sanity input required for the one-hot-plus-zero initialization into the 0 state. As for the four-state diagram, the functions f_{ij}

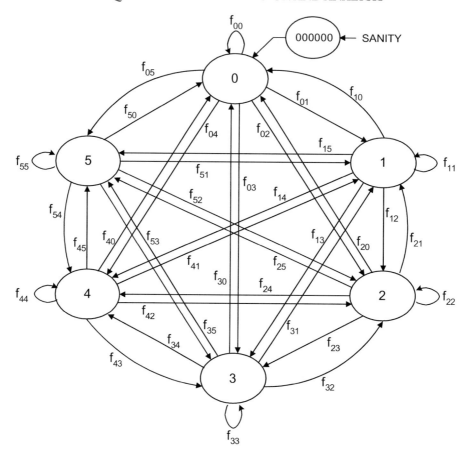

FIGURE 10.4: State diagram for a six-state one-hot sequencer showing the 36 branching function and the one-hot-plus-zero initialization into state 0.

along the leading diagonal of the function matrix represent the holding conditions (six, in this case) each of which is one of six "into" branching functions required for a six-state one-hot sequencer.

Applying Eq. (10.2) to the six-state A-OPS results in the matrix Eq. (10.5), where the "into" terms consist of a 6×6 asymmetric branching function matrix ANDed with a 6×1 matrix of present-state functions, y_i. The first of the "out of" terms results from combining the $y_0 \bar{F}_j$ term to the one-hot-plus-zero term to give $y_0 \bar{y}_1 \bar{y}_2 \bar{y}_3 \bar{y}_4 \bar{y}_5 + \bar{y}_0 \bar{y}_1 \bar{y}_2 \bar{y}_3 \bar{y}_4 \bar{y}_5 = \bar{y}_1 \bar{y}_2 \bar{y}_3 \bar{y}_4 \bar{y}_5$ following application of the factoring and absorptive laws.

$$
\begin{bmatrix} Y_0 \\ Y_1 \\ Y_2 \\ Y_3 \\ Y_4 \\ Y_5 \end{bmatrix}
=
\underbrace{\begin{bmatrix}
f_{00} & f_{10} & f_{20} & f_{30} & f_{40} & f_{50} \\
f_{01} & f_{11} & f_{21} & f_{31} & f_{41} & f_{51} \\
f_{02} & f_{12} & f_{22} & f_{32} & f_{42} & f_{52} \\
f_{03} & f_{13} & f_{23} & f_{33} & f_{43} & f_{53} \\
f_{04} & f_{14} & f_{24} & f_{34} & f_{44} & f_{54} \\
f_{05} & f_{15} & f_{25} & f_{35} & f_{45} & f_{55}
\end{bmatrix}}_{\text{``Into'' Terms}}
\cdot
\begin{bmatrix} y_0 \\ y_1 \\ y_2 \\ y_3 \\ y_4 \\ y_5 \end{bmatrix}
+
\underbrace{\begin{bmatrix}
\bar{y}_1\bar{y}_2\bar{y}_3\bar{y}_4\bar{y}_5 \\
y_1\bar{y}_0\bar{y}_2\bar{y}_3\bar{y}_4\bar{y}_5 \\
y_2\bar{y}_0\bar{y}_1\bar{y}_3\bar{y}_4\bar{y}_5 \\
y_3\bar{y}_0\bar{y}_1\bar{y}_2\bar{y}_4\bar{y}_5 \\
y_4\bar{y}_0\bar{y}_1\bar{y}_2\bar{y}_3\bar{y}_5 \\
y_5\bar{y}_0\bar{y}_1\bar{y}_2\bar{y}_3\bar{y}_4
\end{bmatrix}}_{\text{``Out of'' Terms}}
\tag{10.5}
$$

The $S_j, \bar{R}_j$ functions in Eqs. (10.6) for a six-state A-OPS derive directly from Eq. (10.5) by applying the $Y \rightarrow S\bar{R}$ conversion algorithm Eqs. (5.2) previously given. These functions can now be used in the design of pseudo-Muller circuits by using C-elements. Initialization into state 0 via the 000000 state results by connecting Sanity(L) to the CL(L) input inherent in the C-element CMOS as indicated in Figures 1.12 and 1.13. Thus, a 1(L) (a low voltage) applied to the CL(L) of all C-elements results in the initialization into the "0" state. Sanity circuits have been previously described in Section 2.2.1.

$$
\begin{aligned}
S_0 &= f_{10}y_1 + f_{20}y_2 + f_{30}y_3 + f_{40}y_4 + f_{50}y_5 + \overline{y_1 + y_2 + y_3 + y_4 + y_5} \\
\bar{R}_0 &= f_{00} \\
S_1 &= f_{01}y_0 + f_{21}y_2 + f_{31}y_3 + f_{41}y_4 + f_{51}y_5 \\
\bar{R}_1 &= f_{11} + \overline{y_0 + y_2 + y_3 + y_4 + y_5} \\
S_2 &= f_{02}y_0 + f_{12}y_1 + f_{32}y_3 + f_{42}y_4 + f_{52}y_5 \\
\bar{R}_2 &= f_{22} + \overline{y_0 + y_1 + y_3 + y_4 + y_5} \\
S_3 &= f_{03}y_0 + f_{13}y_1 + f_{23}y_2 + f_{43}y_4 + f_{53}y_5 \\
\bar{R}_3 &= f_{33} + \overline{y_0 + y_1 + y_2 + y_4 + y_5} \\
S_4 &= f_{04}y_0 + f_{14}y_1 + f_{24}y_2 + f_{34}y_3 + f_{54}y_5 \\
\bar{R}_4 &= f_{44} + \overline{y_0 + y_1 + y_2 + y_3 + y_5} \\
S_5 &= f_{05}y_0 + f_{15}y_1 + f_{25}y_{25} + f_{35}y_3 + f_{45}y_4 \\
\bar{R}_5 &= f_{55} + \overline{y_0 + y_1 + y_2 + y_3 + y_4}
\end{aligned}
\tag{10.6}
$$

Shown in Figure 10.5 is the C-element implementation of the six-state A-OPS as derived from Eqs. (10.6). Note that the wireless connection feature is used as emphasized in this text. Also observe that NAND/NOR logic is used including the use of both normal and complementary C-elements given in Figures 1.12 and 1.13, respectively. The A-OPS is initialized into state 0 via the all zero state 000000 following the activation of the a Sanity circuit input, SAN(L) = CL(L) = 1(L).

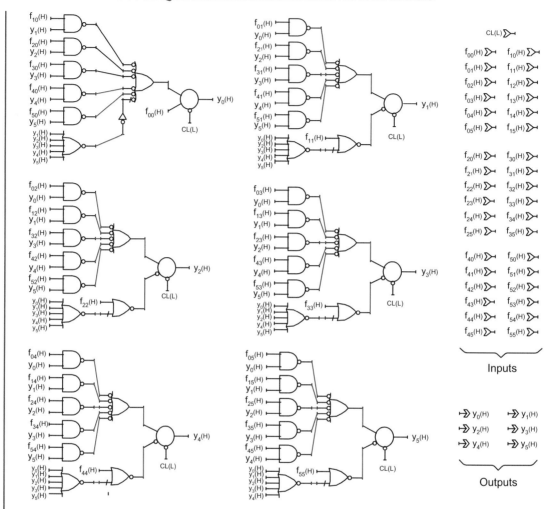

FIGURE 10.5: Simulator macro logic circuit using the wireless connection feature for the six-state A-OPS shown in Figure 10.4 and derived from Eqs. (10.6) as required for implementation with Muller C-elements.

Before the six-state A-OPS is made ready to receive the instruction logic for the time sharing of FSMs, the Sanity circuit must be deactivated, $1(L) \to 0(L)$, and the holding condition for state 0, f_{00}, must be activated $f_{00}(H) \to 1(H)$ and then deactivated. Once stably in state 0, the A-OPS is ready to receive the logic instructions to implement any number of FSMs (e.g., controllers) providing each FSM does not exceed the six-state maximum requirement and is void of all cycle conditions. Having met these requirements, each FSM can be operated on a time-shared basis free of all timing defects. Recall that E-hazards can be activated only if specifically placed delays exceeding a minimum magnitude are present as discussed in Section 5.4. Elimination of E-hazards is easily accomplished by using feedback delays.

10.3 TIME-SHARED MULTIPLE FSM OPERATION BY A SINGLE A-OPS

We will now design three radically different FSMs to be operated via the six-state A-OPS shown in Figure 10.5. To do this, we will follow the generalized architecture given in Figure 10.1 with the PLDs being replaced by three FSMs. Each FSM would, of course, be limited to six states, which is the state limit of the A-OPS. As a necessary requirement, each FSM must be free of cycle and buffer states and, therefore, must have a holding condition for each state. It will be recalled that in the design of one-hot FSMs, state code assignments are unnecessary and are conveniently replaced by state identifiers, numeric or alphabetic. Each FSM will be initialized into the zero state by using the *one-hot plus zero* method. Also, following the emphasis of this text, C-elements will be exclusively used as the memory elements of choice. Because there must be n^2 branching conditions specified for an n-state A-OPS, there will be 36 branching conditions required to be specified for this six-state A-OPS. The three FSMs will be separately activated by enable inputs 0EN, 1EN, and 2EN, which are the outputs from a 2-4 decoder—a time-sharing operation. Each FSM will be represented by an EXL-Sim macro (See Instructional Support Software on p. xiii).

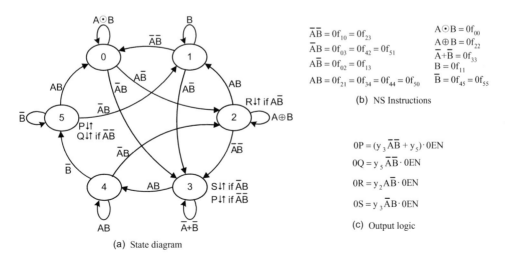

(b) NS Instructions

$$\overline{A}\overline{B} = 0f_{10} = 0f_{23}$$
$$\overline{A}B = 0f_{03} = 0f_{42} = 0f_{51}$$
$$A\overline{B} = 0f_{02} = 0f_{13}$$
$$AB = 0f_{21} = 0f_{34} = 0f_{44} = 0f_{50}$$

$$A\odot B = 0f_{00}$$
$$A\oplus B = 0f_{22}$$
$$\overline{A}+\overline{B} = 0f_{33}$$
$$B = 0f_{11}$$
$$\overline{B} = 0f_{45} = 0f_{55}$$

(c) Output logic

$$0P = (y_3\overline{A}\overline{B} + y_5)\cdot 0EN$$
$$0Q = y_5\overline{A}\overline{B}\cdot 0EN$$
$$0R = y_2A\overline{B}\cdot 0EN$$
$$0S = y_3\overline{A}B\cdot 0EN$$

(a) State diagram

FIGURE 10.6: Design of a six-state FSM, named A-OPS FSM-0, to be implemented by using the six-state A-OPS in Figures 10.4 and 10.5. (a) Fully documented state diagram. (b) NS instructions derived directly from the state diagram. (c) Output logic with enable 0EN.

The first of the three FSMs, to be operated by the six-state A-OPS, is named A-OPS FSM-0 and is defined by the state diagram in Figure 10.6a. It has six states, two input conditions, *A* and *B*, and four outputs, 0*P*, 0*Q*, 0*R*, and 0*S*, where the zero is used to identify these outputs as belonging to FSM-0. The NS instructions and output logic are easily derived directly from the state diagram and are presented in parts (b) and (c) of Figure 10.6. Shown in Figure 10.7 is the macro logic circuit for this FSM generated by the simulator used in this text, EXL-Sim, and briefly discussed on p. xiii. Note that a wireless connection feature is used so as to avoid unnecessary detail.

The second FSM, to be designed for and operated by the six-state A-OPS is named A-OPS FSM-1. It is represented by the fully documented state table in Figure 10.8a. Recall from Section 1.7 that a state table is nothing more than a tabular representation of a state diagram. It is particularly useful in computer-aided design of FSMs.

The NS instructions and output logic are easily extracted directly from the state table and are given in Figure 10.8(b). This FSM has five states, two inputs (*S* and *T*), and two outputs (1*P* and 1*Q*), identifying them as belonging to FSM-1. The EXL-Sim macro logic circuit for this FSM is given in Figure 10.9, which again makes use of the wireless connection feature.

The third FSM designed for and operated via the six-state A-OPS is represented by the fully documented state diagram in Figure 10.10a and is given the name A-OPS FSM-2. It has six states, two external inputs *X* and *Y*, and two outputs, 2*P* and 2*Q*, where the 2's are added to identify these outputs as belonging to FSM-2. The NS instructions and output logic for A-OPS operation are obtained directly from the state diagram and are given in Figure 10.10b and 10.10c. The

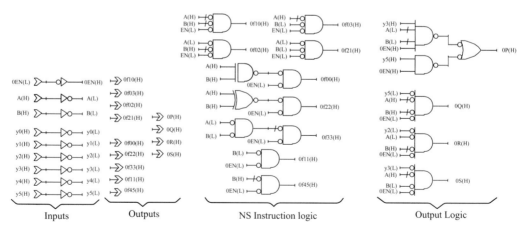

FIGURE 10.7: Simulator macro logic circuit using the wireless connection feature for the FSM named A-OPS FSM-0 derived from Figure 10.6 and to be implemented by using the A-OPS in Figure 10.5.

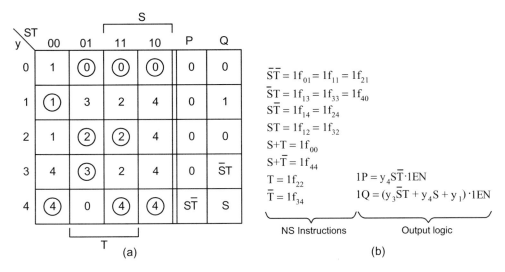

$$\overline{S}\,\overline{T} = 1f_{01} = 1f_{11} = 1f_{21}$$
$$\overline{S}\,T = 1f_{13} = 1f_{33} = 1f_{40}$$
$$S\,\overline{T} = 1f_{14} = 1f_{24}$$
$$S\,T = 1f_{12} = 1f_{32}$$
$$S+T = 1f_{00}$$
$$S+\overline{T} = 1f_{44}$$
$$T = 1f_{22}$$
$$\overline{T} = 1f_{34}$$

$$1P = y_4\,\overline{S}\,\overline{T}\cdot 1EN$$
$$1Q = (y_3\,\overline{S}\,T + y_4\,S + y_1)\cdot 1EN$$

NS Instructions   Output logic

(a)   (b)

FIGURE 10.8: Design of a five-state FSM, named A-OPS FSM-1, to be implemented by using the six-state A-OPS in Figures 10.4 and 10.5. (a) Fully documented state table. (b) NS instructions, and output logic with enable 1EN.

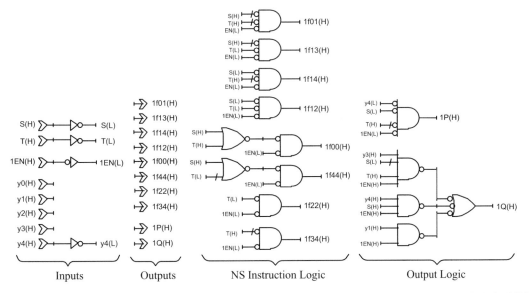

FIGURE 10.9: Simulator macro logic circuit using the wireless connection feature for the FSM A-OPS FSM-1 derived from Figure 10.8 and to be implemented by using the six-state A-OPS in Figure 10.5.

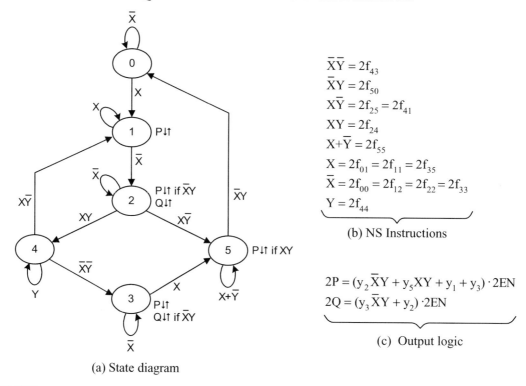

(a) State diagram

$$\overline{X}\overline{Y} = 2f_{43}$$
$$\overline{X}Y = 2f_{50}$$
$$X\overline{Y} = 2f_{25} = 2f_{41}$$
$$XY = 2f_{24}$$
$$X+\overline{Y} = 2f_{55}$$
$$X = 2f_{01} = 2f_{11} = 2f_{35}$$
$$\overline{X} = 2f_{00} = 2f_{12} = 2f_{22} = 2f_{33}$$
$$Y = 2f_{44}$$

(b) NS Instructions

$$2P = (y_2\overline{X}Y + y_5XY + y_1 + y_3) \cdot 2EN$$
$$2Q = (y_3\overline{X}Y + y_2) \cdot 2EN$$

(c) Output logic

FIGURE 10.10: Design of a six-state FSM named A-OPS FSM-2 to be implemented by using the six-state A-OPS in Figures 10.4 and 10.5. (a) Fully documented state diagram. (b) NS instructions. (c) Output logic with enable 2EN.

EXL-Sim macro logic circuit for the NS instructions and output logic is given in Figure 10.11. As with the other logic circuits, a wireless connection feature is used for simplicity.

Note that not all of the branching conditions given in the state diagrams or state table of the three FSMs previously described are represented in their respective macro logic circuits. Only one of the branching conditions for a given set of inputs need be represented in the FSM's macro logic circuit. For example, there are four branching paths associated with input conditions AB in the state diagram of Figure 10.6. We have chosen only the first branching condition $0f_{21}$ to represent the other three, thereby minimizing the logic necessary for that particular macro logic circuit.

Finally, following the generalized A-OPS logic circuit in Figure 10.1, we present in Figure 10.12 the simulator logic circuit for operating the three FSMs on a time-shared basis via the six-state A-OPS. Here, FSM-0, FSM-1, and FSM-2 are all represented as macros together with the macros for the six-state A-OPS and the 2-4 decoder. The interfacing logic is presented by using 27 three-input OR gates but only for the minimum set of branching conditions taken from the logic

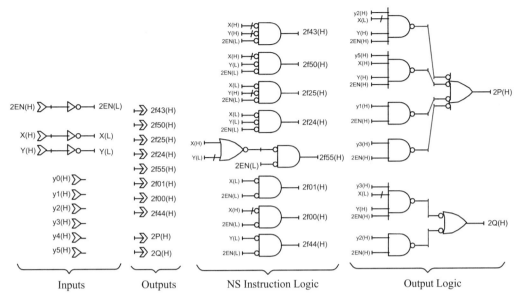

FIGURE 10.11: Simulator macro logic circuit using the wireless connection feature for the FSM A-OPS FSM-2 derived from Figure 10.10 and to be implemented by using the six-state A-OPS in Figure 10.5.

circuits for these FSMs. Obviously, not all of the 36 possible branching paths in the six-state A-OPS of Figure 10.4 are needed for operation of the three FSMs. In fact, there are nine branching paths left unused so that each must be given a forced $0(H)$ input to the six-state A-OPS macro as shown—no input must ever be left dangling. The 2-4 decoder has inputs $I0(H)$ and $I1(H)$ used to select one of the three enables $0EN(L)$, $1EN(L)$, or $2EN(L)$ for operating any one of the three respective FSMs on a time-shared basis. A $CL(L)$ signal from a SANITY circuit initializes to $0(H)$ the six y-variable outputs from the six-state A-OPS, and initializes to $0(L)$ the three outputs from the 2-4 decoder. Exercise: Explain why the missing input to $f11(H)$ in Figure 10.12 (b) must be $1f01(H)$

The mixed-logic simulation of the macro logic circuit in Figure 10.12 is shown in Figure 10.13. In this simulation, three radically different FSMs are operated on a time-shared basis by use of the six-state A-OPS given in Figures 10.4 and 10.5. To do this, a 2-4 decoder serves to enable one of the three FSMs to be operated independently of the other two. Note that all external inputs, A, B, S, T, X, Y, are continuously active throughout the simulation, but only those inputs to an enabled FSM produce the required y variables and external outputs. Each active FSM operates free of endless cycles, critical races, output race glitches, and static hazards. Recall that the internally initiated static hazards formed in the NS functions are automatically covered by the required holding

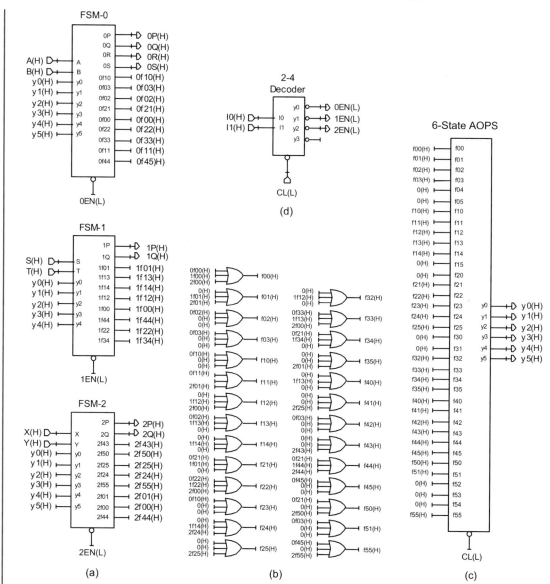

FIGURE 10.12: Simulator macro logic circuit demonstrating the time-shared operation of three radically different FSMs by using an asynchronous one-hot sequencer and the wireless connection feature. (a) Macros for FSM-0, FSM-1 and FSM-2 defined in Figures 10.6–10.11. (b) Interface logic as required by the architecture in Figure 10.1. (c) Macro for the six-state A-OPS in Figure 10.5. (d) The 2-4 decoder used to enable in turn any one of the three FSMs on a time-shared basis.

condition for each state. Essential hazards are possible, but remain only as potential timing defects unless unintended delays exceeding minimum values occur in specific locations within the circuit. Essential hazards in one-hot FSMs always occur via ANDing race gates and are highly predictable but easily eliminated as discussed in Section 5.4.

We have opted to simulate only three FSMs via the A-OPS system. Actually, any number of the FSMs can be operated with a suitable A-OPS but the FSMs are each limited to the maximum number of states established by the A-OPS. The FSMs can themselves be asynchronous system controllers or discrete FSMs. Also, FSMs can be a mixture of both synchronous and asynchronous machines (one-hot designs permit this).

Note in Figure 10.13 that we have chosen to use a CL(L) pulse at each decoder enable to demonstrate that this as an alternative to not using the CL(L) pulse as in Figure 9.18. Either simulation operates perfectly regardless of whether such a pulse is used. However, it does eliminate any possible minute bleed-over of the state variables and outputs during the transitions between FSMs as can be seen in Figure 9.18 although barely detectable.

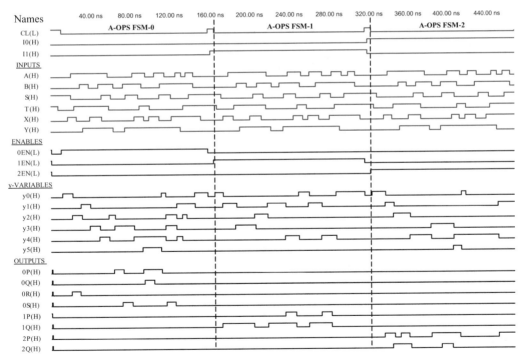

FIGURE 10.13: Simulation of Figure 10.12 showing the sequential behavior of three radically different FSMs defined in Figures 10.6–10.11 on a time-shared basis and operated by using the six-state A-OPS given in Figure 10.5.

10.4 A-OPS SOFTWARE CAPABILITIES USED IN THIS TEXT

The A-OPS CAD software used in this text has the following capabilities. (Note that reference to Muller designs implies the use of C-elements in quasi-Muller circuits.):

1. A-OPS software is applicable to systems up to 12 states. A 12-state A-OPS requires 144 inputs.
2. It can generate a p-term table in the Berkeley format for Muller or Huffman A-OPS designs. These p-term tables are suitable for programming PLAs or programmable array logics (PALs) directly but can be altered to program most any complex PLD.
3. It can generate the VHSIC hardware description language (VHDL) code for C-element designs of quasi-Muller circuits.
4. It can generate the VHDL code for any Huffman A-OPS design.
5. It can generate all the essential hazard (E-hazard) paths and their corresponding ANDing race gates making corrective action easy to eliminate any E-hazard formation.

For the sake of completeness, three sample A-OPS CAD software files are included below showing different output representations for the six-state A-OPS kernel used in this section. The sample CAD software output tables are as follows: Table A-OPS 1 gives the PLA/PAL p-term program table in Berkeley format for a Huffman circuit design; Table A-OPS 2 presents the same information in VHDL code; Table A-OPS 3 provides the $S, \bar{R}$ VHDL code for a Muller circuit design using C-elements.

TABLE A-OPS 1: PLA/PAL P-term program table in Berkeley format for the 6-state A-ops kernel

```
.i 43

.o 6

.ilb y5 y4 y3 y2 y1 y0 f55 f45 f35 f25 f15 f05 f54 f44 f34 f24 f14 f04
f53 f43 f33 f23 f13 f03 f52 f42 f32 f22 f12 f02 f51 f41 f31 f21 f11 f01
f50 f40 f30 f20 f10 f00 sanity
.ob Y5 Y4 Y3 Y2 Y1 Y0
.p 43

-----1-----------------------------------10 000001

----1------------------------------------1-0 000001

---1-------------------------------------1--0 000001
```

TABLE A-OPS 1: (*Continued*)

```
-1--------------------------------------1----0 000001
--1-------------------------------------1---0 000001
1---------------------------------------1-----0 000001
000001-----------------------------------0 000001
-----1----------------------------------1-----0 000010
----1-----------------------------------1-------0 000010
---1------------------------------------1--------0 000010
--1-------------------------------------1---------0 000010
-1--------------------------------------1----------0 000010
1---------------------------------------1-----------0 000010
000010------------------------------------0 000010
-----1------------------1-------------------0 001000
----1-------------------1-------------------0 001000
---1--------------------1-------------------0 001000
--1---------------------1-------------------0 001000
-1----------------------1-------------------0 001000
1-----------------------1-------------------0 001000
000100-------------------------------------0 000100
-----1--------------------1-----------------0 001000
----1---------------------1-----------------0 001000
---1----------------------1-----------------0 001000
--1-----------------------1-----------------0 001000
-1------------------------1-----------------0 001000
1-------------------------1-----------------0 001000
010000-------------------------------------0 010000
-----1-----1--------------------------------0 100000
----1-----1---------------------------------0 100000
---1-----1----------------------------------0 100000
--1-----1-----------------------------------0 100000
-1-----1------------------------------------0 100000
1-----1-------------------------------------0 100000
```

TABLE A-OPS 1: (*Continued*)
100000----------------------------------0 100000
000000----------------------------------0 000001
.e

TABLE A-OPS 2: VHDL code for the six-state A-OPS kernel as required for Huffman circuit designs

```
-- Computer generated VHDL file for a 6 states Async. One-Hot Kernel
-- Kernel with Y output design
-- A-OPS Computer Aided Design Tool
--
library IEEE;
use IEEE.STD_LOGIC_1164.ALL;
use IEEE.STD_LOGIC_ARITH.ALL;
use IEEE.STD_LOGIC_UNSIGNED.ALL;

entity kernel is
      port(
            f5, f4, f3, f2, f1, f0 : in std_logic_vector(5 downto 0);
            sanity: in std_logic;
            Y_O : out std_logic_vector(5 downto 0)
      );
end entity;

architecture structure of kernel is
      signal y, y_l: std_logic_vector(5 downto 0);
      -- Delay definitions
            constant tAND      :time := 3 ns;
            constant tOR :time := 3 ns;
            constant tNOT      :time := 1 ns;
begin
      -- Internal inverted feedback.
      y_l <= not(y) after tNOT;
```

TABLE A-OPS 2: (*Continued*)

```
      y(0) <= ( (f0(0) and y(0)) or (f1(0) and y(1)) or (f2(0) and
y(2)) or (f3(0) and y(3)) or (f4(0) and y(4)) or (f5(0) and y(5))
or (y(0) and y_1(1) and y_1(2) and y_1(3) and y_1(4) and y_1(5)) or
(y_1(0) and y_1(1) and y_1(2) and y_1(3) and y_1(4) and y_1(5))) and
sanity after (tAND+4 ns + tOR);

      y(1) <= ( (f0(1) and y(0)) or (f1(1) and y(1)) or (f2(1) and
y(2)) or (f3(1) and y(3)) or (f4(1) and y(4)) or (f5(1) and y(5)) or
(y_1(0) and y(1) and y_1(2) and y_1(3) and y_1(4) and y_1(5))) and
sanity after (tAND+4 ns + tOR);

      y(2) <= ( (f0(2) and y(0)) or (f1(2) and y(1)) or (f2(2) and
y(2)) or (f3(2) and y(3)) or (f4(2) and y(4)) or (f5(2) and y(5)) or
(y_1(0) and y_1(1) and y(2) and y_1(3) and y_1(4) and y_1(5))) and
sanity after (tAND+4 ns + tOR);

      y(3) <= ( (f0(3) and y(0)) or (f1(3) and y(1)) or (f2(3) and
y(2)) or (f3(3) and y(3)) or (f4(3) and y(4)) or (f5(3) and y(5)) or
(y_1(0) and y_1(1) and y_1(2) and y(3) and y_1(4) and y_1(5))) and
sanity after (tAND+4 ns + tOR);

      y(4) <= ( (f0(4) and y(0)) or (f1(4) and y(1)) or (f2(4) and
y(2)) or (f3(4) and y(3)) or (f4(4) and y(4)) or (f5(4) and y(5)) or
(y_1(0) and y_1(1) and y_1(2) and y_1(3) and y(4) and y_1(5))) and
sanity after (tAND+4 ns + tOR);

      y(5) <= ( (f0(5) and y(0)) or (f1(5) and y(1)) or (f2(5) and
y(2)) or (f3(5) and y(3)) or (f4(5) and y(4)) or (f5(5) and y(5)) or
(y_1(0) and y_1(1) and y_1(2) and y_1(3) and y_1(4) and y(5))) and
sanity after (tAND+4 ns + tOR);

      Y_O <= y;      -- Update output signal with internal feedback.
end structure;
```

TABLE A-OPS 3: VHDL code for the six-state A-OPS kernel as required
for quasi-Muller circuit designs using C-elements

```
-- Computer generated VHDL file for a 6 states Async. One-Hot Kernel
-- Kernel with SR output for C-element design
-- A-OPS Computer Aided Design Tool
--
library IEEE;
use IEEE.STD_LOGIC_1164.ALL;
use IEEE.STD_LOGIC_ARITH.ALL;
use IEEE.STD_LOGIC_UNSIGNED.ALL;

entity kernel is
      port(
             f5, f4, f3, f2, f1, f0 : in std_logic_vector(5 downto 0);
             sanity: in std_logic;
             y : in std_logic_vector(5 downto 0);
             S_O : out std_logic_vector(5 downto 0);
             nR_O : out std_logic_vector(5 downto 0)
      );
end entity;

architecture structure of kernel is
      signal y_l: std_logic_vector(5 downto 0);
      -- Delay definitions
             constant tAND       :time := 3 ns;
             constant tOR :time := 3 ns;
             constant tNOT       :time := 1 ns;
begin
      -- Internal inverted feedback.
      y_l <= not(y) after tNOT;
```

TABLE A-OPS 3: (*Continued*)

```
    S_O(0) <= ( (f1(0) and y(1)) or (f2(0) and y(2)) or (f3(0) and
y(3)) or (f4(0) and y(4)) or (f5(0) and y(5))) and not(sanity);
    S_O(1) <= ( (f0(1) and y(0)) or (f2(1) and y(2)) or (f3(1) and
y(3)) or (f4(1) and y(4)) or (f5(1) and y(5))) and not(sanity);
    S_O(2) <= ( (f0(2) and y(0)) or (f1(2) and y(1)) or (f3(2) and
y(3)) or (f4(2) and y(4)) or (f5(2) and y(5))) and not(sanity);
    S_O(3) <= ( (f0(3) and y(0)) or (f1(3) and y(1)) or (f2(3) and
y(2)) or (f4(3) and y(4)) or (f5(3) and y(5))) and not(sanity);

    S_O(4) <= ( (f0(4) and y(0)) or (f1(4) and y(1)) or (f2(4) and
y(2)) or (f3(4) and y(3)) or (f5(4) and y(5))) and not(sanity);

    S_O(5) <= ( (f0(5) and y(0)) or (f1(5) and y(1)) or (f2(5) and
y(2)) or (f3(5) and y(3)) or (f4(5) and y(4))) and not(sanity);

    nR_O(0) <= ( (f0(0) or (y_l(1) and y_l(2) and y_l(3) and y_l(4)
and y_l(5)) and not(sanity);
    nR_O(1) <= ( (f1(1) or (y_l(0) and y_l(2) and y_l(3) and y_l(4)
and y_l(5)) and not(sanity);
    nR_O(2) <= ( (f2(2) or (y_l(0) and y_l(1) and y_l(3) and y_l(4)
and y_l(5)) and not(sanity);
    nR_O(3) <= ( (f3(3) or (y_l(0) and y_l(1) and y_l(2) and y_l(4)
and y_l(5)) and not(sanity);
    nR_O(4) <= ( (f4(4) or (y_l(0) and y_l(1) and y_l(2) and y_l(3)
and y_l(5)) and not(sanity);
    nR_O(5) <= ( (f5(5) or (y_l(0) and y_l(1) and y_l(2) and y_l(3)
and y_l(4)) and not(sanity);
end structure;
```

CHAPTER 11

Arbiter Modules

The main function of an arbiter is to protect a system from competing inputs. Thus, when two or more inputs are competing for access to a given system, it is the function of an arbiter to arbitrate and grant first access to only one of the competing inputs. This is especially important in Huffman circuits that operate in the fundamental mode. It is for this reason that we opt to design quasi-Muller circuits by using C-elements in the design of most asynchronous sequential circuits including arbiters. C-elements operate outside the fundamental mode, thereby minimizing the probability of metastability developing in a given circuit as discussed in Sections 3.7. Use of a *metastable detection stage* (MDS) in an arbiter is a necessary function to ensure that the signals issued from a metastable condition resolve into either a clean set or reset. This is discussed in Section 8.4. In general, arbiter modules are two-input asynchronous machines but can be combined to accommodate multiple inputs of three or more. We will select a few representative but different arbiters to illustrate the arbitration operation.

11.1 BUS ARBITER MODULE

Many applications of the pulse mode approach to asynchronous finite state machine (FSM) design, discussed in Chapter 6, are prohibited because of overlapping input waveforms. This is a severely limiting requirement of pulse mode FSMs, where nonoverlapping input pulses are mandatory. The bus arbiter module is ideally suited to meet this requirement. A bus arbiter arbitrates two request inputs on a *first-in/first-out* basis but grants access to the second request only after the first request goes inactive (is withdrawn). If both request inputs arrive simultaneously, the *mutual exclusion* (ME) character of the bus arbiter must arbitrate access to only one request input at a time—a formidable task considering that metastability may also become a factor. The use of C-elements and a *metastability detection stage* in the design of the bus arbiter help to significantly mitigate this problem. It is true that the MDS also has an inherent ME feature that we combine as a single stage designated as metastable detection/mutual exclusion (MD/ME) stage.

Shown in Figure 11.1 are the state diagrams, output logic, logic circuit, and circuit symbol for the C-element design of the bus arbiter. Included is the MD/ME stage following Section 8.4, whose function it is to detect and eliminate any metastable condition developed in either C-element and to issue a clean set or clean reset to only one output grant signal at a time. The design converts overlapping requests, R_1, R_2, to a nonoverlapping grant signal G_1 or G_2 on the basis of the arbitration and mutually exclusivity function of the arbiter. The results are clean mutually exclusive and minimally separated grant output pulses suitable for use in pulse mode design applications.

The simulation of the bus arbiter shown in Figure 11.1 is given in Figure 11.2. Here, the primary function of the bus arbiter module is clearly indicated by the output responses to a given random set of input pulses some of which are overlapping. For the propagation delays set for the various components of the circuit, the grant pulses (G_1, G_2) are minimally separated by the delay through an inverter. Also, any individual, discrete R pulse (see simulation >300 ns) strong enough to cross the switching threshold, will generate a grant pulse that is never less than the path delay of an inverter. Notice that the function $R_0(L)$ goes active anytime both grant pulses are simultaneously inactive. Thus, $R_0(L)$ pulses are indicative of the number of grant pulses that have been issued, a useful design parameter for some applications. Finally, observe that when $CL(L)$ is active no grant pulse will be issued even as the request input signals continue.

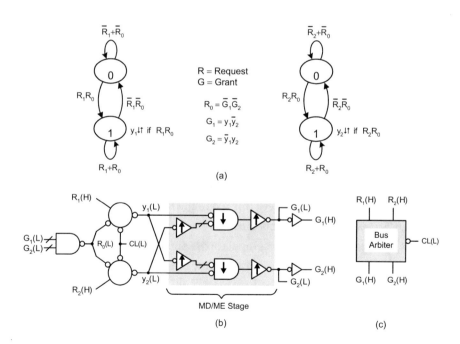

FIGURE 11.1: Design of the bus arbiter module. (a) State diagrams and output logic for the two C-elements. (b) Logic circuit using the wireless connection feature for C-elements with active low outputs and the MD/ME stage discussed in Section 8.4. (c) Circuit symbol.

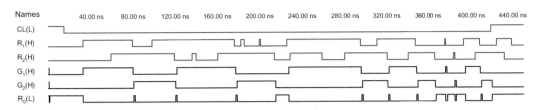

FIGURE 11.2: EXL-Sim simulation of the bus arbiter logic circuit in Figure 11.1b showing the basic operation of the bus arbiter together with the effect of a CLEAR (CL).

11.2 MULTIPLE INPUT BUS ARBITERS

Typically, non-handshake arbiter modules have two request inputs and two grant outputs, and can vary considerably in design form and function. The bus arbiter, discussed in Section 11.1, is a typical example of a two-input/two-output arbiter. As we shall see, such arbiters can be combined with *rendezvous modules* (RMODs) to yield multiple input arbiter configurations. Thus, in this section, we will limit our discussion to multiple input bus arbiters. However, we do recognize that there are alternative designs for multiple input arbiters. We will focus our attention on two very different examples in succeeding chapters.

To arbitrate *n-inputs* to a protected system requires N_n arbiters taken q at a time as given by Eq. (11.1), which requires n C-elements (RMODs), each with $(n-1)$ inputs. For bus arbiters $q = 2$ (inputs).

$$N_n = \frac{n!}{q!(n-q)!} = \text{No. of arbiters} \qquad (11.1)$$

As used in arbiters, a C-element is called an RMOD.

A three-input bus arbiter requires $N_n = 3!/2!(3-2)! = 3$ arbiters each with $(n-1) = 2$ inputs and three 2-input RMODs. Shown in Figure 11.3 are the block diagram and the circuit symbol for such an arbiter. As stated earlier, this arbiter operates on a first-in/first-out basis, but thereafter prioritizes the grant output selection following withdrawal of the first request. Further grant output selection is by priority following withdrawal of the two previous requests.

The simulation of this three-input bus arbiter is shown in Figure 11.4 following a CL(L) signal change $1(L) \to 0(L)$. Note the random overlapping input request signals and the resulting discrete nonoverlapping grant output signals that reveal the required first-in/first-out priority grant-select behavior. Activation of the CL(L) to 1(L) removes all output grant signals while allowing the input request signals to remain.

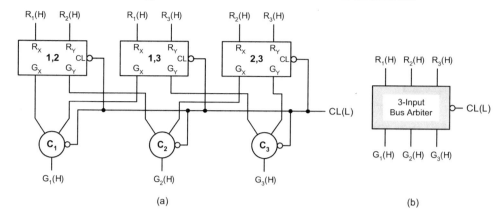

FIGURE 11.3: (a) Three-input bus arbiter with CLEAR by using C-element RMODs. (b) Circuit symbol.

A variety of multiple input bus arbiters are possible by following Eq. (11.1). Listed in Table 11.1 are six possibilities ranging from three to eight inputs showing the number of arbiters, RMODs, and RMOD inputs required for each. For example a four-input arbiter, numbered 3, 2, 1, 0, would require six arbiters to be connected in the following pairs—32, 31, 30, 21, 20, and 10—requiring four RMODs each with three inputs. To accommodate the application of three or more RMODs requires use of C-element trees as indicated in Figure 11.5a, 11.5b, and 11.5c. At some point, say more than four inputs, it might be advantageous to use the generalized hybrid form in Figure 11.5d or the complementary metal-oxide semiconductor alternative to (d) in Figure 11.5e.

A short exercise: Discuss whether the generalized C-element in Figure 11.5e can be used to implement any one of the p-terms in Eqs. (9.7). Support your answer by an example using mixed-logic notation in Figure 11.5e.

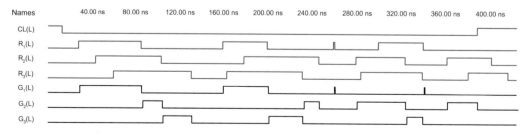

FIGURE 11.4: Simulation of the logic circuit for the three-input bus arbiter in Figure 11.3 showing the first-in/first-out priority grant-select behavior of the arbiter and the effect of CLEAR.

TABLE 11.1: Requirements for multiple input arbiters with $q = 2$ by using Eq. (11.1).

INPUTS, n	ARBITERS, N_n	NO. OF RMODS	RMOD INPUTS
3	3	3	2
4	6	4	3
5	10	5	4
6	15	6	5
7	21	7	6
8	28	8	7

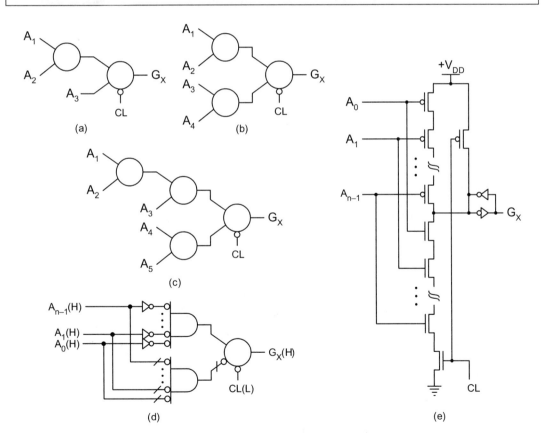

FIGURE 11.5: RMODs for use with multiple-input arbiters. (a) Asymmetric three-input C-element tree. (b) Symmetric four-input C-element tree. (c) Asymmetric five-input C-element tree. (d) Generalized hybrid NOR/INV RMOD . (e) Generalized CMOS alternative to the hybrid form in (d).

11.3   PRIORITY STAND-ALONE ARBITERS

Arbiters that have a fixed number of allowed inputs and cannot be combined to increase the number of its inputs are called *stand-alone* arbiters. Shown in Figure 11.6a is the state diagram for a three-input stand-alone arbiter. The entered variable (EV) K-maps and optimum cover for the LPD design of this arbiter are given in Figure 11.6b, from which Eqs. (11.2) results. Here, we use the $Y \rightarrow S\bar{R}$ conversion algorithm because the state diagram in Figure 11.6a is interpreted as having a one-hot state code assignment. The C-element logic circuit in Figure 11.6c derives from Eqs. (11.2)

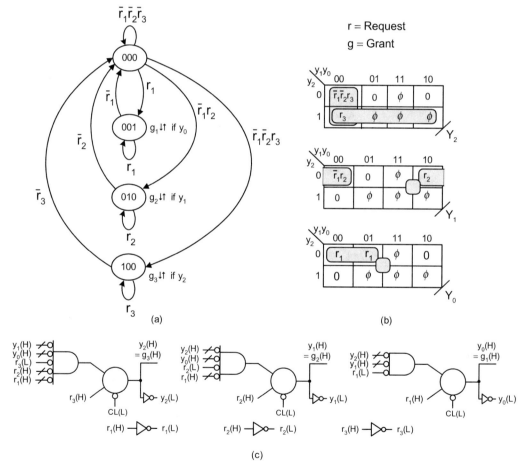

(a)

(b)

(c)

FIGURE 11.6: Design of the three-input stand-alone priority arbiter . (a) Fully documented state diagram for minimum input and output logic. (b) NS K-maps. (c) C-element Logic circuits using the wireless connection feature for simplicity and following Eqs. (11.2) for the $Y \rightarrow S,\bar{R}$ conversion algorithm given by Eqs. (5.2).

and is initialized into the 000 state by using the CL(L) input to the C-elements—no need for the one-hot-plus-zero method.

$$
\begin{aligned}
Y_2 &= \bar{y}_1\bar{y}_0 r_3\bar{r}_2\bar{r}_1 + y_2 r_3 &\rightarrow\quad S_2 &= \bar{y}_1\bar{y}_0 r_3\bar{r}_2\bar{r}_1 & \bar{R}_2 &= r_3 \\
Y_1 &= \bar{y}_2\bar{y}_0 r_2\bar{r}_1 + y_1 r_2 &\rightarrow\quad S_1 &= \bar{y}_2\bar{y}_0 r_2\bar{r}_1 & \bar{R}_1 &= r_2 \\
Y_0 &= \bar{y}_2\bar{y}_1 r_1 + \bar{y}_0 r_1 &\rightarrow\quad S_0 &= \bar{y}_2\bar{y}_1 r_1 & \bar{R}_0 &= r_1 \\
g_1 &= y_0 \qquad g_2 = y_1 & g_3 &= y_2
\end{aligned}
\qquad (11.2)
$$

The simulation of the stand-alone priority arbiter in Figure 11.6 is shown in Figure 11.7 with and without overlapping input request waveforms. It is easy to see that the operation of this arbiter follows the requirements established by the state diagram in Figure 11.6a. Hence, note that the output grant signals, $g_i = y_i$, never overlap, making this arbiter suitable for use with pulse mode circuit designs.

11.4 HANDSHAKE ARBITERS WITH ACKNOWLEDGMENT (DONE) SIGNALS

Figure 11.8 shows the state diagrams, and the C-element implementation for the handshake arbiter module with an MD/ME stage. This arbiter module is the most versatile of all the arbiter modules. For example, it can be used as bus arbiter module by setting $D_X(H) = D_Y(H) = 0(H) = 1(L)$. Or with each Grant signal connected to its corresponding Done input, it can be combined with other such arbiter modules to produce a multiple-input bus arbiter as in Figure 11.3. Furthermore, the Done signals can be delivered back to the handshake arbiter modules from different parts of the circuit the arbiter is protecting. Finally, as will be demonstrated in Section 11.4, the handshake arbiter module together with Trans-HI and Trans-LO modules will be used in a rotating token arbiter. It is important to consider that in all these different usages of the handshake arbiter module, ME, and metastability protection is guaranteed by virtue of the MD/ME stages. This fact becomes important when it is remembered that C-elements operate outside the fundamental mode but can still go

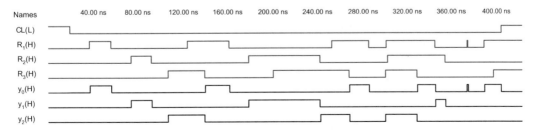

FIGURE 11.7: Simulation of the three-input stand-alone priority arbiter circuit in Figure 11.6c showing its basic operation as designed with C-elements.

metastable as discussed in Section 3.7. Also, refer to Section 8.4 for a discussion of the MDS. There, we show how MDS detects metastability and protects its outputs from passing on that metastability to the inputs of any FSM or combinational logic stage to which the arbiter's output are connected. For this system, the mean time between failures (MTBF) can be raised to extremely high values, but not to infinity.

The state diagrams in Figure 11.8a are those representing the normal C-element shown in Figure 1.12. However, the C-elements in the logic circuit of Figure 11.8c are complementary but used as normal C-elements with activation levels for $Q_X(L)$ and $Q_Y(L)$ matching those of the C-elements. For a better understanding of what was done here and for the various alternative C-element configurations available for use, refer to Figure 2.2.

The simulation of the handshake arbiter module is presented in Figure 11.9. Here, the dependence of the Grant outputs are seen to depend on the random set of Done signals received from the protected system following a random set of Request signals from the source system. Generally, a given Request will be granted on a first-in/first-out basis, but only if its corresponding Done signal

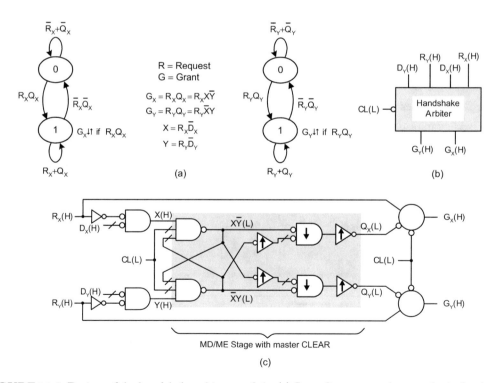

FIGURE 11.8: Design of the handshake arbiter module. (a) State diagrams and output logic for the two C-elements. (b) Circuit symbol. (c) Logic circuit with request and done (acknowledgement) inputs and grant outputs and highlighting the MD/ME stage that accommodates a master CLEAR.

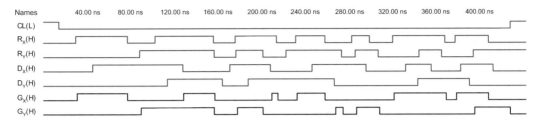

FIGURE 11.9: Simulation of the handshake arbiter circuit in Figure 11.8 showing the dependence of the Grant outputs on random Done (acknowledgment) inputs received from the protected system following a given Request from the source system.

is low $0(H)$ or goes low after the Request has been issued from the source. However, once a request has been granted, the corresponding Done signal can go active $1(H)$ without affecting the Grant signal that was previously granted. If the two Request signals go active at the same time each with low Done signals, then arbitration must select one and grant the request. If a Done signal is then issued for the selected Request, the other Request will be granted as shown in Figure 11.9.

11.5 ROTATING TOKEN ARBITERS

We present here an unusual multiple input arbiter that is characterized by a rotating signal, called a token. This arbiter is used to arbitrate on the basis of this token signal but with some interesting characteristics. The *rotating token arbiter* module uses the handshake arbiter in Figure 11.8 as its center but combines with *Trans-HI* and *Trans-LO* modules also known as transparent D-Latches—the word "Trans" means *Transparent* because the data D input transfers through the latch when CK is active high (H) or active low (L), respectively. Before moving on to the arbiter module architecture, it is useful to define the Trans-Hi and Trans-LO modules. Shown in Figure 11.10a and 11.10b are the state diagrams and EV K-maps for these two modules. The K-maps are plotted by using the excitation table in Figure 1.14d for the combined basic SR cells and following the mapping algorithm given in Section 1.6. The minimum cover is indicated by the shaded loops with the results provided to the right of each K-map. The same result would have been obtained by first mapping using the lumped path delay (LPD) model for the y variables and then converting to the SR by using the $Y \rightarrow SR$ map conversion algorithm in Section 2.1. The C-element logic circuits and circuit symbols are shown in Figure 11.10c and 11.10d for the Trans-HI and Trans-LO modules.

Simulation of the Trans-HI and Trans-LO modules is provided in Figure 11.11. Here, rising edge triggering, falling edge triggering, and the transparent character are each is shown. Remember that the waveform for CK(L) is the logic inverse of that for CK(H). Thus, the reader should mentally invert the CK(L) waveform to confirm that it indeed produces falling edge triggering. In short,

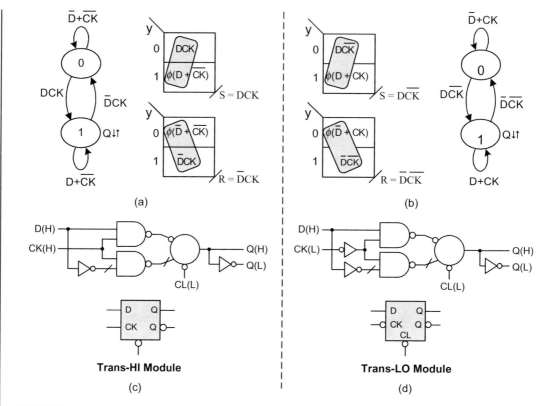

FIGURE 11.10: C-element design of the Trans-HI and Trans-LO modules. (a and b) State diagram, SR K-maps and SR logic for the Trans-HI and Trans-LO modules. (c and d) Logic circuit and circuit symbol for the Trans-HI and Trans-LO modules.

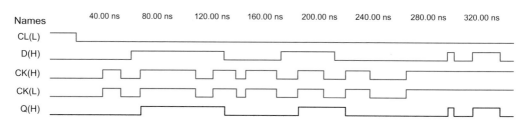

FIGURE 11.11: Simulation of the Trans-HI and Trans-LO modules in Figure 11.10 showing $0 \rightarrow 1$ transitions and transparency of $Q(H)$ when $CK(H)$ or $CK(L)$ are active.

if the D input to these modules is presented to the Trans-HI or Trans-LO module at the time CK is active, it will be directly transferred to the output. Otherwise, the D input must be clocked to the output on a $0 \rightarrow 1$ transition of CK(H) or CK(L).

The block diagram logic circuit for the *rotating token arbiter module* is given in Figure 11.12(a) with its circuit symbol provided in Figure 11.12(b). The operation of the rotating token arbiter follows that of the handshake described in Section 11.3 but with the inclusion of the Trans-HI and Trans-LO modules and XOR gate. Basically, an active Grant signal $G_X(H)$ is issued following a request $R_X(H)$ provided that $D_x(H)$ is inactive ($0(H)$), and that T_{in} and T_{out} are of opposite activation levels. This, in turn, causes the CK(H) to the Trans-HI module to go active , if only briefly, thereby issuing a $G_X(H)$ output. The reader should follow the simulation of Figure 11.12 presented in Figure 11.13.

In Figure 11.14, the rotating token arbiter module is used in a series configuration of the first four stages in an n-stage (hence, n-input) rotating token arbiter, where the Done signal inputs originate from any desirable part of the protected system or systems. The rotating token is initiated by a $START \oplus T_{n+1}$ signal from the XOR gate. A three-stage simulation of the rotating token arbiter is given in Figure 11.15, where the Done signals have been purposely delayed by different amounts to produce a more realistic simulation. Note that the rotating token T_i signals are paused intermittently due to the manner in which the Done signals are received by the arbiter. As a result, some of the Request signals fail to produce a Grant output. In fact, any Grant output $G_i(H)$ can be selectively made inactive by holding a Done signal $D_i(H) = 1(H) = 0(L)$ for some arbitrary time but

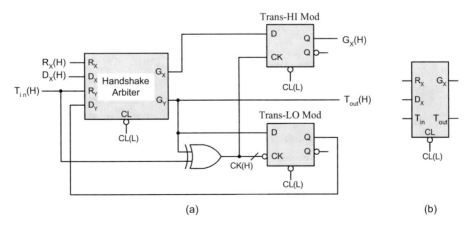

(a)   (b)

FIGURE 11.12: Design of the rotating token arbiter module by using the handshake arbiter and Trans-HI and Trans-LO modules. (a) Logic circuit. (b) Circuit symbol.

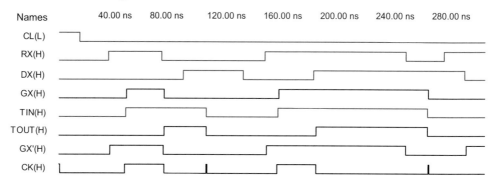

FIGURE 11.13: Simulation of the rotating token arbiter module in Figure 11.12 showing the Request, Done and Grant relationships with the Done (acknowledgement) signal.

only prior to the rising edge of the $R_i(H)$ Request input. For comparison purposes, another simulation of this three-stage arbiter in provided in Figure 11.16 where, in this case, the Grant signal for each stage is fed back into the Done input of that stage without delay and with no significant pauses in the rotating token signals. Thus, $D_i(H) = G_i(H)$, where every Request signal produces a Grant signal of similar duration but delayed to an extent depending on the relative position of the rotating token.

11.6 APPLICATIONS

Having read the contents of this chapter, the reader should be left with the impression that a multitude of arbiter applications are available to the logic designer, actually too many to discuss in detail

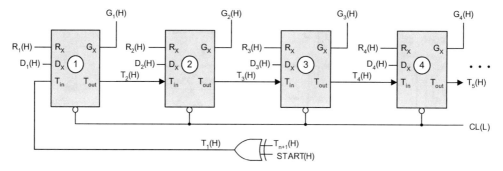

FIGURE 11.14: The first four stages of an n-stage rotating token arbiter initiated with a START $\oplus$ T_{n+1} signal from the XOR gate and with the DONE signals taken from any desirable part of the protected system.

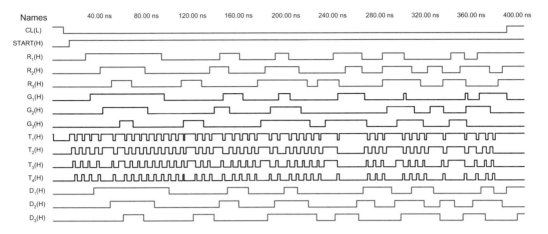

FIGURE 11.15: Simulation of the three stages of the rotating token arbiter in Figure 11.14 showing the dependence of the Grant signals on the Requests, Done acknowledgements and the rotating token signals.

given the limited space provided in the text. In this section, we will touch on some applications of the rotating token arbiter and then present in some detail an important application of the bus arbiters.

One interesting application of the rotating token arbiter concept is to activate and coordinate the outputs from different FSMs. In this way, outputs of one FSM could be issued conditional on

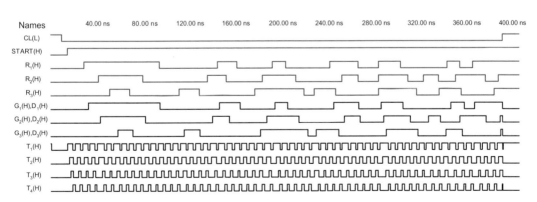

FIGURE 11.16: For comparison with Figure 11.15, a simulation of Figure 11.14 is made with the same Request and Done signals but now with the Grant signals of each stage connected to the Done input of that stage.

the Done signals from another FSM. Clearly, each Request input to a given FSM must correspond to the Grant signal from that FSM, but regulated by a Done signal from the same or from some other FSM. Multiple Grant outputs from each FSM could be issued dependently or independently of the Done signals issued by the rotating token arbiters of other FSMs. A type of pipelining operation can be obtained if the Grant outputs are fed into the Request inputs of the succeeding stages with or without outputs to the external world. An array of rotating token arbiters with the appropriate Request, Grant, and Done signal interconnects could conceivably be used in complex pipelining operations. The master CLEAR signal for each rotating token arbiter can be used to pause the entire arbiter operation at any time. Alternatively, any individual Done signal can be used to selectively deactivate its respective Grant output for some predetermined period. Furthermore, the START signal for each arbiter in the array can be used to arrest the rotating token for that arbiter leaving only the Request and Done signals active. Overall, the possibilities here are nearly unlimited. Again, keep in mind that each rotating token arbiter module is designed with a built-in metastable detection system. If a metastable condition should occur within one of the arbiter stages, its operation would pause until the metastable condition is resolved to a clean set or reset.

The pulse mode approach offers a simple and reliable means of designing asynchronous FSMs free of the many timing defects uniquely owned by clock-independent systems. The characteristics

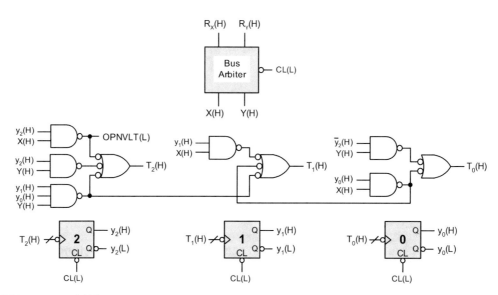

FIGURE 11.17: (a) The pulse-mode 2-bit digital combinational lock in Figures 6.6 and 6.7 now configured with the bus arbiter of Figure 11.1 used to convert overlapping input waveforms to nonoverlapping discrete pulse signals as required by all pulse mode operations.

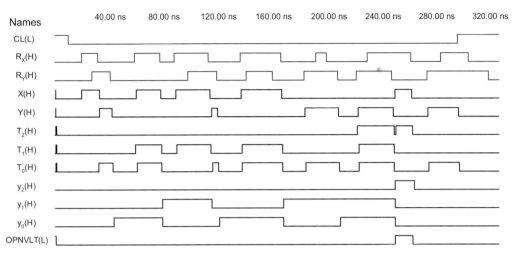

FIGURE 11.18: Simulation of the pulse-mode 2-bit combinational lock in Figures 6.6 and 6.7 by using the bus arbiter in Figure 11.1 to deal with overlapping input waveforms.

of the pulse mode approach are detailed in Section 6.1.1. From this, it is clear that the one important drawback to designing in the pulse mode is the severe limitation of having discrete nonoverlapping data input signals. Use of the bus arbiter is ideally suited to overcoming this limitation. Shown in Figure 11.17 is the logic circuit for the design of the pulse-mode 2-bit digital combinational lock of Figures 6.6 and 6.7, but now with its two inputs taken from the bus arbiter featured in Figure 11.1 and simulated in Figure 11.2.

Simulation of the pulse mode digital combinational lock in Figure 11.17 is shown in Figure 11.18. Note the transitions from overlapping pulses to discrete nonoverlapping pulses. Although some of the edges of these transitions appear to be very close, they are actually separated by an inverter path delay. Further separation is easily produced by logic with larger path delays or by inserting delays within the bus arbiter. A simple means of introducing delays is to place pairs of inverters in the appropriate places. The obvious place for such delays in the bus arbiter is on the Grant output lines.

APPENDIX A

Brief Reviews

A.1 MIXED-LOGIC GATE SYMBOLOGY AND CONJUGATE GATE FORMS

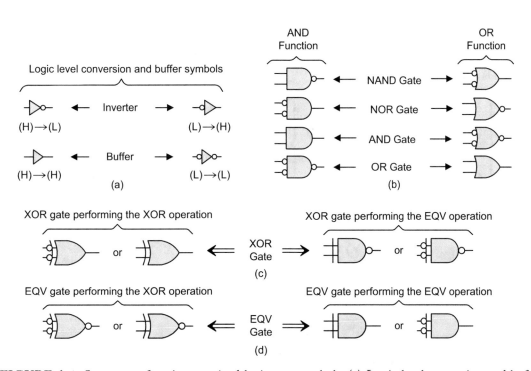

FIGURE A.1: Summary of conjugate mixed logic gate symbols. (a) Logic level conversion and buffer symbols. (b) AND and OR gate symbology. (c) and (d) XOR and EQV gate symbology and logic operations.

A.2 AND/OR LAWS AND THE EQV/XOR LAWS OF BOOLEAN ALGEBRA (DUAL RELATIONS)

(Note: Symbols $\odot = \overline{\oplus}$ both represent equivalence, EQV)

Associative laws

$$\left\{\begin{array}{l} (X \cdot Y) \cdot Z = X \cdot (Y \cdot Z) = X \cdot Y \cdot Z \\ (X + Y) + Z = X + (Y + Z) = X + Y + Z \end{array}\right\}$$

$$\left\{\begin{array}{l} (X \overline{\oplus} Y) \overline{\oplus} Z = X \overline{\oplus} (Y \overline{\oplus} Z) = X \overline{\oplus} Y \overline{\oplus} Z \\ (X \oplus Y) \oplus Z = X \oplus (Y \oplus Z) = X \oplus Y \oplus Z \end{array}\right\} \quad (A.1)$$

Commutative laws

$$\left\{\begin{array}{l} X \cdot Y \cdot Z = X \cdot Z \cdot Y = Z \cdot X \cdot Y = \cdots \\ X + Y + Z = X + Z + Y = Z + X + Y = \cdots \end{array}\right\}$$

$$\left\{\begin{array}{l} X \overline{\oplus} Y \overline{\oplus} Z = X \overline{\oplus} Z \overline{\oplus} Y = Z \overline{\oplus} X \overline{\oplus} Y = \cdots \\ X \oplus Y \oplus Z = X \oplus Z \oplus Y = Z \oplus X \oplus Y = \cdots \end{array}\right\} \quad (A.2)$$

Distributive laws

$$\left\{\begin{array}{l} \text{Factoring law} \\ (X \cdot Y) + (X \cdot Z) = X \cdot (Y + Z) \\ \text{"Distributive law"} \\ (X + Y) \cdot (X + Z) = X + (Y \cdot Z) \end{array}\right\} \quad \left\{\begin{array}{l} \text{Factoring law} \\ (X \cdot Y) \oplus (X \cdot Z) = X \cdot (Y \oplus Z) \\ \text{"Distributive law"} \\ (X + Y) \overline{\oplus} (X + Z) = X + (Y \overline{\oplus} Z) \end{array}\right\} \quad (A.3)$$

Absorptive laws

$$\left\{\begin{array}{l} X \cdot (\overline{X} + Y) = X \cdot Y \\ X + (\overline{X} \cdot Y) = X + Y \end{array}\right\} \quad \left\{\begin{array}{l} X \cdot (\overline{X} \oplus Y) = X \cdot Y \\ X + (\overline{X} \overline{\oplus} Y) = X + Y \end{array}\right\} \quad (A.4)$$

Consensus laws

$$\left\{\begin{array}{l} XY + \overline{X}Z + (YZ) = (XY + \overline{X}Z) \\ (X + Y)(\overline{X} + Z)(Y + Z) = (X + Y)(\overline{X} + Z) \end{array}\right\}$$

$$\left\{\begin{array}{l} XY \oplus \overline{X}Z + (YZ) = (XY \oplus \overline{X}Z) \\ (X + Y) \overline{\oplus} (\overline{X} + Z)(Y + Z) = (X + Y) \overline{\oplus} (\overline{X} + Z) \end{array}\right\} \quad (A.5)$$

Useful Identities	
AND/OR LAWS	**XOR/EQV LAWS**
$X \cdot 0 = 0$	$\bar{X} \oplus Y = X \oplus \bar{Y} = \overline{X \odot Y} = \bar{X} \odot \bar{Y} = X \oplus Y$
$X \cdot 1 = X$	$\bar{X} \odot Y = X \odot \bar{Y} = \overline{X \oplus Y} = \bar{X} \oplus \bar{Y} = X \odot Y$
$X \cdot X = X$	$X \odot \bar{X} = X \oplus X = 1$
$X \cdot \bar{X} = 0$	$X \oplus \bar{X} = X \oplus X = 0$
$X + 0 = X$	$X \oplus 1 = \bar{X}$
$X + 1 = 1$	$X \oplus 0 = X$
$X + X = X$	$X \odot 1 = X$
$X + \bar{X} = 1$	$X \odot 0 = \bar{X}$
$X \oplus Y \equiv X\bar{Y} + \bar{X}Y$	$X \odot Y \equiv \bar{X} \cdot \bar{Y} + X \cdot Y$

For functions α and β: If $\alpha \cdot \beta = 0$ then $\alpha \oplus \beta = \alpha + \beta$.

For functions α and β: If $\alpha + \beta = 1$ then $\alpha \odot \beta = \alpha \cdot \beta$.

DeMorgan's laws (complementation)

$$\left\{ \begin{array}{l} \overline{X \cdot Y} = \bar{X} + \bar{Y} \\ \overline{X + Y} = \bar{X} \cdot \bar{Y} \end{array} \right\} \quad \left\{ \begin{array}{l} \overline{X \odot Y} = \bar{X} \oplus \bar{Y} = X \oplus Y \\ \overline{X \oplus Y} = \bar{X} \odot \bar{Y} = X \odot Y \end{array} \right\} \tag{A.6}$$

A.3   ENTERED VARIABLE K-MAP COMPRESSION AND MINIMIZATION

As a first example, consider the conventional 1's and 0's K-map and function Y expressed as the sum of minterms in Figure A.2a. Next, compress function Y into a second-order K-map with A as the entered variable (EV), and extract minimum cover as in Figure A.2b. Next, compress function Y into a first-order K-map (with EVs A and C) in Figure A.2c and extract minimum cover with $A \oplus C = A\bar{C} + \bar{A}C$. Notice the OPIs in Figure A.2c.

A.3.1 Incompletely Specified Functions

ϕ = don't care and can be taken as either a logic 0 or logic 1, take your choice. A don't care function, say ϕX, represents an *incompletely specified function* or one that is *nonessential*. Any function α ANDed

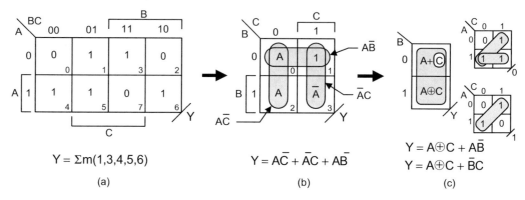

FIGURE A.2: (a) Conventional third-order K-map for function Y. (b) First-order compression in a second-order EV K-map with entered variable A showing minimum cover for Y. (c) Second-order compression in a first-order EV K-map with A and C as EVs and showing use of the adjacent XOR pattern in (b) to give the alternative minimum cover in (c) together with diagonal XOR pattern in the first order conventional submaps for A and C in the B domains.

with ϕ, hence $\phi\alpha$, means that function α is nonessential and can be used in K-map minimization as either $0 \times \alpha = 0$ or $1 \times \alpha = \alpha$ as needed. See Glossary.

Rules to remember

Use incompletely specified functions in K-map minimization so as to achieve optimum logic cover by following three rules.

1. Treat the don't care (ϕ) as an EV, which it is because it can be taken as a logic 0, a logic 1, or an input. Remember: *Use it if you can; otherwise, ignore it.*

2. In simplifying incompletely specified functions, simplify by applying the *absorptive* laws:

$$X + \phi\bar{X} = X + \phi \quad \text{and} \quad X \cdot (\phi + \bar{X}) = \phi X \tag{A.7}$$

where now $X + \phi$ can be taken as X or 1 depending on its best usage. The term ϕX can be taken as either an X or a 0. Note: Functions of the type $X + \phi$ have an essential sum-of-products component, X or 1. What about functions of the type ϕX?

3. Terms such as $\phi(X + Y)$ often occur. This means that $(X + Y)$ is nonessential. To use such terms effectively, simplify using the factoring law as

$$\phi(X + Y) = \phi X + \phi Y \tag{A.8}$$

then choose how best to use, or not to use, this result in a K-map minimization operation.

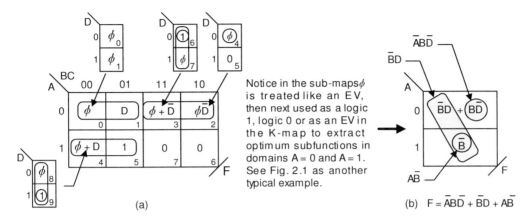

Notice in the sub-maps ϕ is treated like an EV, then next used as a logic 1, logic 0 or as an EV in the K-map to extract optimum subfunctions in domains A = 0 and A = 1. See Fig. 2.1 as another typical example.

(a)

(b) $F = \overline{AB}\overline{D} + \overline{B}\overline{D} + A\overline{B}$

FIGURE A.3: Minimization of the function F in Eq. (A.9). (a) First-order compression in A, B and C with entered variable D, and showing four first-order sub-maps with minterm code numbers corresponding to those in Eq. (A.9). (b) Third-order compression in A with EVs B and D showing minimum cover indicated by the shaded loops (independent of input C).

Example

Given the following incompletely specified function represented in Minterm code (see Glossary):

$$F(A, B, C, D) = \sum m(3, 6, 9, 10, 11) + \phi(0\ 1, 4, 7, 8) \tag{A.9}$$

To obtain an optimum cover for function F, first compress it into a third-order K-map in A, B, and C with EV D, then into a first-order K-map in A with EVs B and D as in Figure A.3. For a first-order compression, count by 2's in the minterm code of Eq. (A.9) beginning with (0,1) and enter the results into first-order submaps. Then enter each submap result into the third-order K-map as in Figure A.3a.

APPENDIX B

End-of-Chapter Problems

CHAPTER 1

1.1 Invert the state code assignments in the three-state finite state machine (FSM) given in Figure 1.6a and do the following:

   (a) Construct the state table for this FSM with the altered state code assignments.

   (b) Construct the second-order entered variable (EV) K-maps for next state (NS) variables Y_1 and Y_0, and for output Z.

   (c) Loop out a minimum cover from these K-maps. *Partial answer*: $Y_0 = A\bar{B} + y_1\bar{B} + y_0A$.

   (d) Construct the logic circuit from the results of (c) exclusive of fictitious memory elements.

CHAPTER 2

2.1 (a) Repeat Problem 1.1(b) with the altered state code assignments and use the K-map conversion algorithm for S_1, R_1 and S_0, R_0 as given in Section 2.1.

   (b) Loop out a minimum cover from these K-maps. *Partial answer*: $S_1 = A$, $R_0 = \bar{A}B$.

   (c) Construct the logic circuit assuming the use of normal C-elements with CL(L) and simulate the result by initializing into state 00.

   (d) Construct the logic circuit assuming the use of Set dominant basic cells as the memory and simulate the result by initializing into state 11.

   (e) Do the simulation results of (c) and (d) compare? Explain.

   (f) Compare the results of part (b) with those of Eqs. (2.1). What do you conclude?

2.2 Redesign both basic cells in Figures 1.8 and 1.9 by using C-elements. Simulate and compare the results with those in Figure 1.10. *Partial answer*: $R_{C\text{-element}} = \bar{S}R$ for set-dominated basic cell.

CHAPTER 3

3.1 The state diagram in Figure 11.10a is that for a Trans-HI module otherwise known as an RET transparent D-latch.

(a) Design this FSM by using the lumped path delay (LPD) model and sum-of-products (SOP) logic. Show that it contains an externally initiated static-1 hazard, indicate how it is formed and give its required hazard cover.

(b) Simulate this LPD circuit with and without the hazard cover.

3.2 Shown in Figure B.1 is the state diagram of an FSM to be analyzed for static hazards.

(a) Design this FSM by using the LPD model by obtaining the NS SOP expressions for Y_1 and Y_0. Show that this FSM contains an internally initiated static hazard. Show how this hazard is formed and find its hazard cover. *Partial answer*: $Y_1 = \bar{y}_0 \bar{A} B + y_1 B + y_1 \bar{y}_0$.

(b) Simulate this FSM with and without the hazard cover.

(c) By using the $Y \rightarrow SR$ K-map conversion algorithm described in Section 2.1, obtain the Set–Reset (SR) logic required for implementation of a nested C-element or nested cell design. Determine whether this SR design contains a static hazard.

3.3 The FSM in Figure 3.10 contains an output race glitch (ORG) but no static hazard.

(a) Explain how this ORG can form.

(b) Correct the state diagram so as to make the formation of an ORG impossible. (Hint: The transition from 11 to 00 must pass through two states.)

(c) Redesign this FSM for the LPD model and use the $Y \rightarrow SR$ conversion algorithm to obtain the SR logic suitable for a C-element design. Compare these results with those of Eqs. (3.3).

(d) Check for static hazards in the new results of (c) and if they exist, eliminate them.

(e) Simulate the results of part (c) to prove that it follows the corrected state diagram of (b).

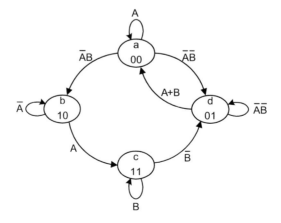

FIGURE B.1: An FSM to be analyzed for static hazards.

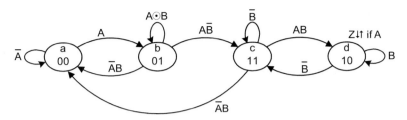

FIGURE B.2: State diagram of an FSM that possesses a critical race that must be eliminated.

3.4 Shown in Figure B.2 is a four-state FSM that contains a critical race. If the critical race is not eliminated, it can cause the FSM to malfunction.

(a) Explain how the critical race can form and under what conditions.

(b) Correct the state diagram in Figure B.2 to show that the critical race is not possible.

(c) With the corrected state diagram of part (b), design this FSM by using the LPD model followed by the $Y \rightarrow SR$ conversion algorithm in Section 2.1 to obtain the SR logic suitable for using C-elements as the memory.

(d) Simulate the results of (c) and verify that it follows the corrected state diagram.

3.5 Shown in Figure B.3 is an FSM that has both a potential essential hazard (E-hazard) and a potential d-trio. (Note: This is an advanced exercise requiring a complete understanding of Section 3.6.)

(a) By using the LPD model, find the NS functions, Y_1 and Y_0, and note that there are no static hazards present.

(b) Read Section 3.6 carefully, then run a complete E-hazard and d-trio analysis on this FSM. To do this, follow Section 3.6.1 and the example in Section 3.6.2. Find the

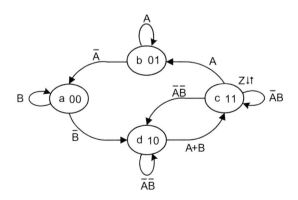

FIGURE B.3: An FSM that contains an E-hazard and a d-trio

input initiation requirements, the initiator input, and the first and second invariants for each. Determine the direct and indirect paths for E-hazard and d-trio formation, their race gates (ANDing or ORing), and the minimum delays, Δt_E and Δt_D, required to activate each, respectively.

(c) Repeat parts (a) and (b) for an SR C-element design of this FSM by using the $Y \rightarrow SR$ conversion algorithm in Section 2.1.

(d) Test the LPD and SR results of (a)–(c) by simulation with and without Δt_E and Δt_D.

CHAPTER 4

4.1 The FSM in Figure 3.10 is to be designed by using the *single time transition (STT) array algebraic approach*. To do this, use the alphabetic state identifiers, follow Sections 4.1 and 4.2, and do the following:

(a) Construct the state table from the state diagram for this FSM.

(b) Find the π- and τ-partitions, and the state matrix **S** and the destination matrix **D**. *Partial answer*: $\pi_1 = a,bc = \tau_1$.

(c) From the results of (b) find the function matrix F_{NS} and the state functions Y_1 and Y_0. *Partial answer*: $Y_0 = y_0\bar{A} + y_1\bar{B} + \bar{A}B$. (Hint: There is an externally initiated static hazard in Y_1.) Compare these results with those of Eqs. (3.3). What do you conclude?

(d) Following a similar procedure as in (b) and (c), find the output logic for Z and show that no ORG exists. Note that this result is not the same as in Eq. (3.6). In what way do these results differ and has the sense of the sequential nature of this FSM been altered by the STT approach? Explain.

(e) Will an E-hazard analysis of the STT results differ from those given in Section 3.6.2? Explain.

4.2 Shown in Figure B.4 is a five-state, two-input and three-output FSM that is to be designed via the STT array algebraic method. The state diagram and state table for this FSM are given in parts (a) and (b). Note that the state table is unfolded in Gray code as has been the practice up to this point.

Given in Figure B.5 are the π- and τ-partitions, one possible state assignment matrix **S**, which are derived from Figure B.4, and the state assignment map, which is presented as a fourth-order EV K-map. Now, by following the example in Section 4.2, complete the design of this FSM by doing the following:

(a) Obtain the destination matrix from the state table in Figure B.4.

(b) Determine from Figure B.5 and Section 4.2 how many valid **S** matrices are possible.

(c) From the results of (a), find the function matrix F_{NS} making use of the state assignment map in Figure B.5c to determine that $e = y_4$, $bc = \bar{y}_1y_0$, $d = y_1y_0$, $abcde = 1$, among

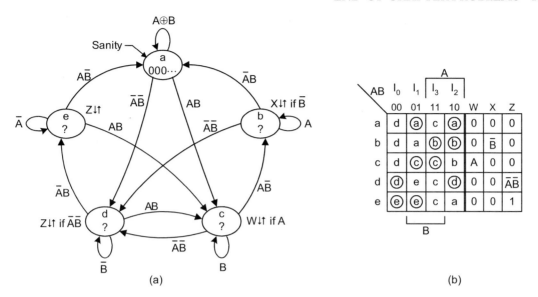

(a)   (b)

FIGURE B.4: A five-state, two-input, three-output FSM to be design by using the STT array algebraic method. (a) State diagram. (b) State table derived from (a) laid out in Gray code.

others. Now, find the NS functions for Y_4, Y_3, Y_2, Y_1, Y_0. Use the laws of Boolean algebra to simplify the expressions wherever possible. Check for static hazards and eliminate any that exist. *Partial answer*: $Y_1 = \bar{A}\bar{B} + y_1\bar{A} + y_1 y_0 \bar{B}$.

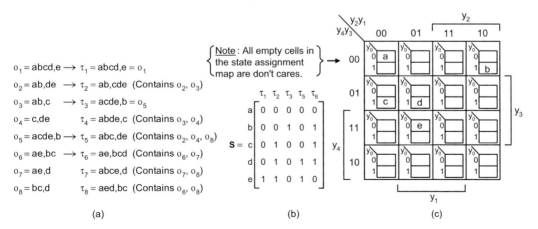

(a)   (b)   (c)

FIGURE B.5: (a) π-partitions, and a minimum set of τ-partitions. (b) State assignment matrix obtained from the τ-partitions. (c) Fourth-order EV state assignment map obtained from the state assignment matrix, **S**.

(d)   Determine the outputs W, X, and Z, and again simplify wherever possible.

(e)   Study the state diagram in Figure B.4 and determine how many E-hazard paths satisfy the minimum requirements given in Section 3.6.1. Give the path for each including their initiating input conditions and race gates (ANDing or ORing). Hint: There
are more than four E-hazard paths.

4.3 Design the FSM in Figure B.4 by using the CAD program ADAM. To do this, review
Section 4.4 and carefully read the Readme.doc that accompanies the software. Remember
that the state table must be unfolded in binary *not* in Gray code and it must not include the
output functions. Three separate tables are required for the three outputs. Use a .txt editor
for both the input file and the batch file. (Note: This is an advanced exercise requiring a
complete understanding of the ADAM software.)

(a)   Run ADAM and find the π- and τ-partitions, the state assignment matrix, the desti-
nation matrix. Also find the cubes for the NS functions Y_4, Y_3, Y_2, Y_1, Y_0, and for the
SR equations as required for a C-element design. Lastly, find the p-term table in the
Berkeley format for a PLA design of this FSM.

(b)   Check for static hazards and eliminate any that exist in the LPD results.

(c)   Discuss the inherent differences between the "pencil-and-paper" FSM design in
Problem 4.2 and that of the ADAM-CAD design of this problem (4.3). In particular,
discuss the relative use of shared PIs by these two approaches.

CHAPTER 5

5.1 In Figure B.6a is a four-state asynchronous FSM, which has two inputs, X and Y, and two
outputs, P and Q, that is to be designed via the one-hot-plus-zero method. After reading
Sections 5.1 and 5.2, follow the design example in Section 5.3 and do the following:

(a)   Write directly from the state diagram the NS and output expressions assuming that
the FSM has been initialized into the 0000 state by the one-hot-plus-zero method.
Simplify the expression for Y_a by applying the factoring and absorptive laws. *Partial
answers*: $Y_b = bX\bar{Y} + b\bar{X}Y + a\bar{X}Y + d\bar{X}Y + b\bar{c}\bar{d}$ and $P = b\bar{X} + c\bar{X}Y$.

(b)   From the results of (a), construct the logic circuit assuming the LPD model. Use the
wireless connection feature wherever convenient. Note that static hazards are not possible in properly designed one-hot FSMs.

(c)   Simulate this FSM and confirm that each one-hot state-to-state transition must pass
through a state having two 1's—one "1" for the initial state and the other "1" for the
destination state.

(d)   Read Section 5.4 and analyze any E-hazards this FSM may have. (Hint: There are two,
assuming that $\bar{X}Y \rightarrow X\bar{Y}$ in state b is not permitted, since it can create a function hazard.)

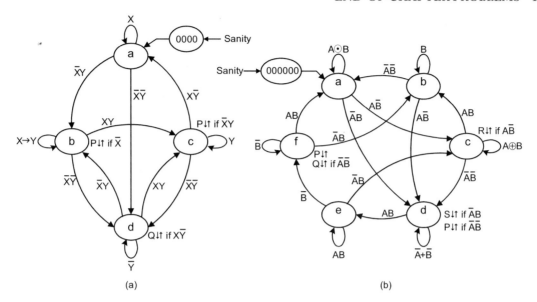

FIGURE B.6: Two asynchronous FSMs that are to be designed by using the one-hot-plus-zero method. (a) A four-state FSM with two inputs and two outputs. (b) A six-state FSM with two inputs and four outputs.

5.2 Repeat parts (a) and (b) of Problem 5.1 for the six-state asynchronous FSM in Figure B.6b, but now use the one-hot $Y \rightarrow S\bar{R}$ conversion algorithm in Section 5.2 to design with C-elements. *Partial answers:* $S_a = b\bar{A}\bar{B} + fAB + \bar{b}\bar{c}\bar{d}\,\bar{e}\bar{f}$; $R_d = \bar{A} + \bar{B} + \bar{e} = \overline{ABe}$.

  (a) Simulate the C-element design of this FSM and confirm that the state diagram in Figure B.6b is followed.

  (b) Determine how many potential E-hazards in this FSM satisfy the minimum requirements for E-hazard formation and give the initiating conditions for each. For example, an input change of $AB \rightarrow A\bar{B}$ in state a causes the transition $a \rightarrow c \rightarrow b \rightarrow d$. (Hint: There are more than nine E-hazards.)

CHAPTER 6

6.1 Design the FSM in Figure B.7a as a pulse mode machine by using falling edge-triggered (FET) toggle modules as the memory. To do this, first read Sections 6.1, 6.2, and 6.3.

  (a) By using the "T" excitation table in Figure 6.5, plot the two second-order K-maps with EVs X and Y, then extract minimum cover for the two NS functions T_1 and T_0, and for the output Z.

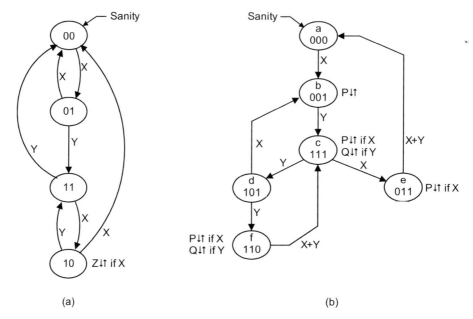

FIGURE B.7: Two FSMs to be design in the pulse mode by using toggle modules. (a) A four-state sequence recognizer with two inputs and one output. (b) A six-state FSM with two inputs and two outputs.

    (b)   Construct the logic circuit by using FET toggle modules as the memory following Figures. 6.3 and 6.4. The use of toggle module macros is highly recommended.

    (c)   Simulate the circuit and verify that it conforms to the state diagram in Figure B.7a.

  6.2 Repeat parts (a), (b), and (c) for the design of the pulse-mode FSM in Figure B.7b. To do this in part (a), plot three third-order NS K-maps and two output K-maps with EVs X and Y. Now complete parts (b) and (c) as in Problem 6.1.

CHAPTER 7

  7.1 A logic circuit is represented by three NS functions and two output functions given below that have been read directly from the logic circuit. Analyze this FSM by doing the following with help from Sections 7.1 and 7.2.

$$Y_2 = y_2 \bar{A} \bar{B} + A\bar{B}$$
$$Y_1 = y_1 \bar{A} \bar{B} + \bar{A}B + y_1 y_0 AB$$
$$Y_0 = y_2 \bar{A} \bar{B} + \bar{A}B + y_0 AB + A\bar{B}$$
$$X = y_2 \bar{A} \bar{B} + y_1 y_0 AB$$
$$Z = y_1 \bar{A} \bar{B} + \bar{y}_1 y_0 AB$$

(a) Plot the third-order EV K-maps for the NS and output functions, then construct the PS/NS table from which the state diagram can be constructed. Identify all don't care states that exist and show their branching relationship with the primary state routine. Follow Figure 7.6 in this regard.

(b) What kind of FSM is this, based on what has been covered so far in this text? Indicate how this FSM should be initialized.

(c) By using the state diagram, check for the presence of endless cycles, critical races, ORGs, and static hazards. If any of these timing defects are present, show how they can be eliminated. Thus, if static hazards are present, provide the correct hazard cover to eliminate them. (Hint: There are six externally initiated static-1 hazards in this FSM.)

(d) Run an E-hazard analysis and identify any potential E-hazards or d-trios that may exist.

7.2 The NS and output function for an asynchronous FSM shown below are read directly from a pulse mode circuit. From this information, construct the state diagram for this FSM and analyze it.

$$T_2 = \bar{y}_2\bar{y}_1 y_0 Y + y_2 X \quad T_1 = \bar{y}_2 y_1 X + y_0 Y \quad T_0 = \bar{y}_2 y_1 X + \bar{y}_2 y_1 Y + y_2 \bar{y}_0 Y + y_2 \bar{y}_1 Y + \bar{y}_0 X$$

$$P = y_1 X + \bar{y}_2 \bar{y}_1 y_0 \quad Q = y_2 y_1 Y$$

(a) Map these NS and output functions into third-order EV K-maps and perform the $T \rightarrow Y$ K-map conversion given in Section 7.1, part 2b. Note that this conversion follows the XOR relation for $T \rightarrow Y$ giving $Y = y \oplus T = \bar{y}T + y\bar{T}$.

(b) From the results of (a), construct the PS/NS table and the state diagram for this FSM. Verify that it satisfies the requirements for a pulse mode FSM and note any problems it may have. Indicate how the FSM should be initialized.

(c) Can endless cycles, critical races, ORGs, static hazards or E-hazards occur in this pulse mode FSM? Explain.

7.3 The NS and output functions below are those for an FSM design by the one-hot-plus-zero method by using C-elements as the memory.

$$S_a = e\bar{S}T + \bar{b}\bar{c}\bar{d}\bar{e} \qquad \bar{R}_a = S + T + \bar{b}$$

$$S_b = a\bar{S}\bar{T} + c\bar{S}\bar{T} \qquad \bar{R}_b = \bar{S}\bar{T} + \bar{c}\bar{d}\bar{e}$$

$$S_c = bST + dST \qquad \bar{R}_c = T + \bar{b}\bar{e} \qquad P = eS\bar{T}$$

$$S_d = b\bar{S}T \qquad \bar{R}_d = \bar{S}T + \bar{c}\,\bar{e} \qquad Q = d\bar{S}T + eS + b$$

$$S_e = bS\bar{T} + cS\bar{T} + d\bar{T} \qquad \bar{R}_e = S + \bar{T} + \bar{a}$$

Analyze this FSM by doing the following:

(a) Carry out the reverse conversion $S\bar{R} \to Y$, then use the Y functions directly to construct the state diagram. It will be helpful to consult Sections 5.1, 5.2, 5.3, and 7.4 before starting on part (a). *Partial answer:* $Y_b = a\bar{S}\bar{T} + c\bar{S}\bar{T} + b\bar{S}\bar{T} + b\bar{c}\bar{d}\bar{e}$.

(b) Simulate this FSM to show that it conforms to the state diagram.

(c) Follow Section 7.4 and run an E-hazard analysis on this FSM by giving the E-hazard paths, initiating conditions, initiator input, first and second y-variable invariants, direct and indirect paths, race gates (ANDing or ORing), and the minimum path delays required to activate each E-hazard.

CHAPTER 8

8.1 The four-state FSM in Figure B.8a is a simple sequence recognizer having two inputs and a single output. Design this FSM by using the externally asynchronous/internally clocked system (EAIC) method. To accomplish this, do the following:

(a) Following Section 8.2, design and make a macro of the DFLOP shown in Figures 8.3, 8.5, and 8.6.

(b) Simulate the DFFOP design in (a) and verify that it conforms to the state diagrams in Figure 8.3.

(c) From the state diagram in Figure B.8a, plot the D_A and D_B EV K-maps by using the excitation table in Figure 1.2b, the mapping algorithm in Section 1.6, and its application in Sections 1.7 and 8.3.

(d) Extract minimum cover from the EV K-maps in part (c). *Partial answer:* $D_A = A\bar{X}Y + BXY + AB$.

(e) Use the DFLOP shown in Figure 8.6 and design the logic circuit for this FSM following the example in Section 8.3. For simplicity, use the wireless connection feature emphasized in this text.

(f) Simulate the resulting EAIC circuit and demonstrate that the simulation conforms to the state diagram.

8.2 Repeat parts (a), (b), (c), (d), (e), and (f) of Problem 8.1 for the FSM in Figure B.8b. *Partial answer:* $D_C = \bar{A}\bar{S}\bar{T} + CT$.

CHAPTER 9

9.1 (a) Read Sections 9.1 and 9.2 and then, following Section 9.3, design and make a macro of the 2 × 2 microprogrammable asynchronous controller (MAC) module by using C-elements.

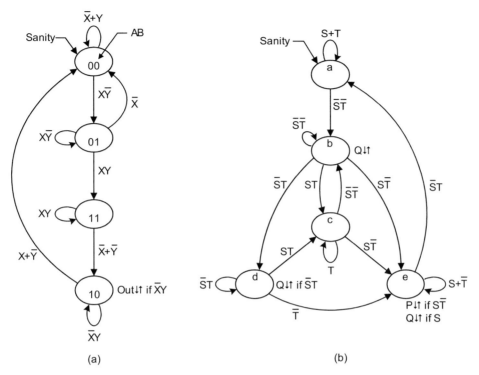

FIGURE B.8: Design of two FSMs by using the EAIC system. (a) A simple four-state sequence recognizar having two inputs and one output. (b) A more complex five-state FSM having two inputs and two outputs.

(b) Test the macro of (a) by simulation to verify that the macro conforms to the state diagrams in Figure 9.3.

(c) Cascade the 2 × 2 MAC module with another to produce a 4 × 4 MAC module as in Figure 9.7 and test it by simulation. Making a macro of the 4 × 4 MAC module is optional.

9.2 Shown in Figure B.9 are three FSMs to be designed by the 4 × 4 MAC module and operated on a time-shared bases. After completing Problem 9.1, do the following:

(a) By following the examples in Section 9.5, construct the MAC program table for each FSM, plot the NS instruction inputs in appropriate EV K-maps, and loop out an optimum cover for each. *Partial answer*: $I_2 = y_2\bar{y}_0\,\bar{S}T + \bar{y}_2 y_1 y_0 S + y_2\bar{y}_1\bar{S}$ for MAC-0.

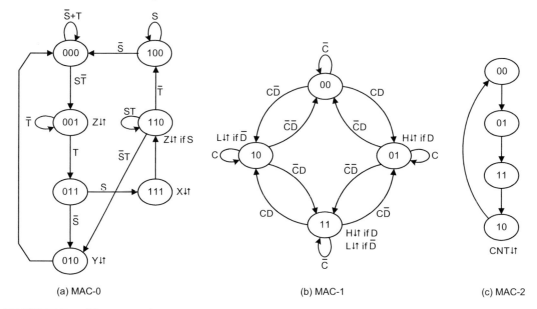

FIGURE B.9: Three asynchronous FSMs to be design by using the 4 × 4 MAC module in Figure 9.7 and operated on a time-shared basis after having been initialized into the all-zero state. (a) MAC-0, a seven-state FSM of two inputs and three outputs. (b) MAC-1, a selector module with two inputs and two outputs. (c) MAC-2, a 2-bit gray code counter with one output.

(b)   Following Section 9.6 and Figures 9.17 and 9.18, simulate the time-shared operation of the three FSMs by using the 4 × 4 MAC module. Take particular caution in setting up the interfacing logic and making the appropriate use of the DI inputs to the MAC-2 FSM.

CHAPTER 10

10.1 (a)   Following Sections 10.1 and 10.2, design and make a macro of the six-state asynchronous one-hot programmable sequencer (A-OPS) shown in Figure 10.4.

(b)   Test the macro by simulation and verify that it conforms to the state diagram in Figure 10.4.

10.2 Shown in Figure B.10 are three FSMs to be designed via the six-state A-OPS in Figure 10.4 and operated by the six-state A-OPS on a time-shared basis. To do this, do the following:

(a)   By following Section 10.3, obtain the NS instruction and output logic for each FSM directly from their respective state diagrams. Refer to Eqs. (10.1) and (10.2) before starting on this exercise.

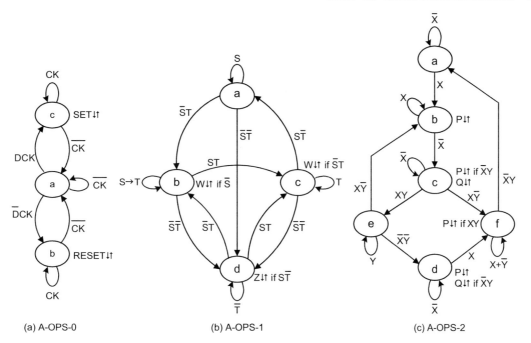

(a) A-OPS-0   (b) A-OPS-1   (c) A-OPS-2

FIGURE B.10: Three asynchronous FSMs to be designed by using the six-state A-OPS in Figure 10.4 and to be operated on a time-shared basis. (a) A three-state FSM with two inputs and two outputs. (b) A four-state FSM with two inputs and two outputs. (c) A six-state FSM with two inputs and two outputs.

(b)   From the results of (a), construct the logic circuit for each FSM. Note: It is advisable to make macros for each of these FSMs. Compare the results of Figure B10b with that of Problem 5.1.

(c)   Simulate the time-shared operation of these three FSMs following the format of Figures 10.12 and 10.13. Note that it is not necessary to pulse the CL(L) input for each enable EN(L) from the decoder.

(d)   Use the A-OPS software to generate all E-hazard paths, their initiators and first- and second-invariants, and the ANDing race gates so that corrective action can easily be taken if necessary.

CHAPTER 11

11.1 (a)   Following Section 11.1, design and make a macro of the bus arbiter shown in Figure 11.1.

(b)   Test the macro by simulation to verify that it conforms to the state diagrams and output logic in Figure 11.1a.

(c)   By following Section 11.1.1, design and simulate a four-input bus arbiter by using C-elements as the RMODs. Verify that the grant outputs do not overlap. To do this, blow up a waveform portion.

11.2 (a)   Use the FET toggle module designed in Section 6.2 and redesign the pulse mode FSM in Figure 7.11 by using the two-input bus arbiter in Problem 11.1 to deal with overlapping inputs R_X and R_Y as in Figure 11.17.

(b)   Simulate the result of part (a) by using similar R_X and R_Y waveforms as in Figure 11.18.

11.3 (a)   Follow Section 11.3 and make a macro of the handshake arbiter in Figure 11.8. Simulate to verify that it conforms to the state diagrams, output logic and circuit in Figure 11.8.

(b)   By following Section 11.4, design and make macros of the Trans-Hi and Trans-LO modules by using C-elements. Simulate to verify that they conform to the state diagrams in Figure 11.10.

(c)   Following Section 11.4, design and make a macro of the rotating token arbiter module. Simulate this arbiter module and compare with Figure 11.13.

(d)   Following Figures 11.14, 11.15, and 11.16 and by using the results of parts (b) and (c), design a four-input rotating token arbiter with grant signals connected back into the respective Done signals. Simulate to verify its proper operation.

Additional Problems and Exercises: Invert state assignments or reassign state identifiers in state diagrams where appropriate and rework the problems in Chapters 1 through 11.

Advanced Projects:
PROJECT I
    Repeat Problems 5.1 and 5.2 by using the A-OPS software and VHDL.
PROJECT II
    Repeat Problems 10.1 and 10.2 by using the A-OPS software and VHDL.

Endnotes

I. GENERAL BACKGROUND DIRECTLY SUPPORTING MATERIAL IN THIS TEXT

Supporting material for nearly all subject matters covered in this text can be found in Chapters 14, 15, and 16 of the revised second edition of Tinder's book cited below. Additional supporting material and associated work by Tinder and others can be found in publications by VanScheik and Tinder, and by Tinder et al.

R. F. Tinder, *Engineering Digital Design*, 2nd ed. Revised, Academic Press, San Diego, CA, 2000.

W. S. VanScheik and R. F. Tinder, "High speed externally asynchronous/internally clocked systems," *IEEE Trans. Comput.* **46**(7), 824–829 (1997).

R. F. Tinder, R. I. Klaus, and J. A. Snodderley, "High speed microgrammable asynchronous controller modules," *IEEE Trans. Comput.* **43**(10), 1226–1232 (1994).

II. ALTERNATIVE APPROACHES TO ASYNCHRONOUS STATE MACHINE DESIGN AND ANALYSIS

Comparatively speaking, the book by Myers offers a significantly different approach to asynchronous finite state machine (FSM) design by using Petri nets, set theory, positive logic, state graphs, burst-mode, and extended burst-mode FSMs, and VHSIC hardware description language, among others features. It also provides discussions of both Huffman- and Muller-type syntheses and emphasizes the use of C-elements. The book by Myers provides an extensive list of 425 references, some of which, however, are only peripherally related to the subject matter of Tinder's text.

C. J. Myers, *Asynchronous Circuit Design*, John Wiley & Sons, New York, 2001.

III. IMPORTANT HISTORICAL CONTRIBUTIONS TO ASYNCHRONOUS CIRCUIT SYNTHESIS

Many publications of the past fall into this category. However, there are a few that stand out as having had a significant impact on the subject matter presented in this text by Tinder. The book by

Unger and the contributions of Huffman, and Muller are good examples. Also, the publication by Maki and Tracey provide an early look at the single transition time approach to asynchronous FSM design.

S. H. Unger, *Asynchronous Sequential Switching Circuits*, Wiley-Interscience, New York, 1969.

D. A. Huffman, "The synthesis of sequential switching circuits," in E. F. Moore, Ed., *Sequential Machines: Selected Papers*, Addison-Wesley, Reading, MA, 1964.

D. E. Muller, "The general synthesis problem for asynchronous digital networks," in *Annual Symposium on Switching and Automata Theory*, New York, 1967.

G. K. Maki and J. H. Tracey, "A state assignment procedure for asynchronous sequential circuits," *IEEE Trans. Comput.* **20**, 666–668, 1971.

IV.   SOURCES RELATED TO THE SUBJECT OF EAIC SYSTEMS DISCUSSED IN THIS TEXT

A variety of systems have been studied that utilize both internally fixed and pausable clocks, but for various reasons are inherently more complex than the externally asynchronous/internally clocked system (EAIC) system described at length in this text. In further contrast, these systems offer little or no protection from metastable effects—an important feature of the EAIC approach. There are two selected references that are typical of these studies. They are works by Nowick and Dill, and by Rosenberger et al. The work of Rosenberger et al. describes the design and analysis of Q-flops in an internally clocked configuration. The Q-flops are designed with an internal handshaking mechanism to ensure that the inputs are not stored until the input stage is ready to accept them, and the outputs are not updated until the input stage has fully resolved and is stable in its new state. This allows the design of sequential delay-insensitive modules that require fewer delay constraints than other functionally equivalent design methodologies.

S. M. Nowick and D. L. Dill, "Automatic synthesis of locally-clocked asynchronous state machines," *Proc. ICCAD*, 1991.

F. U. Rosenberger, C. E. Molnar, T. J. Chaney, and T. Fang, "Q-modules: internally clocked delay-insensitive modules," *IEEE Trans. Comput.* **37**(9), 1005–1018, 1988.

Glossary of Terms, Expressions, and Abbreviations

ABEL   Advanced Boolean expression language.

Activate   To assert or make active.

Activation level   The logic state of a signal designated to be active or inactive.

Activation level indicator   A symbol, (H) or (L), attached to a signal name to indicate positive logic or negative logic, respectively.

Active   A descriptor that denotes an action condition and that implies logic 1.

Active high (H)   A term that indicates a positive logic source or signal.

Active low (L)   A term that indicates a negative logic source or signal.

Active state   The logic 1 state of a logic device.

Active transition point   The point in a voltage waveform where a digital device passes from the inactive state to the active state.

ADAM   Automated design of asynchronous machines.

Adjacent cell   A K-map cell whose coordinates differ from that of another cell by only 1 bit.

Adjacent pattern   An XOR pattern involving an uncomplemented function in one cell of a K-map and the same function complemented in an adjacent cell.

Algorithm   Any special step-by-step procedure for accomplishing a task or solving a problem.

Alternative race path   One of two or more transit paths an FSM can take during a race condition.

ALU   Arithmetic and logic unit.

AMC   Asynchronous microcontroller.

Analog   A term that refers to continuous signals such as voltages and current, in contrast to digital or discrete signals.

Analysis of FSMs   The procedure that yields a state diagram, or state table, beginning with a logic circuit or its VHDL representation. Also the procedure used to identify and eliminate timing defects in asynchronous FSMs.

AND   An operator requiring that all inputs to an AND logic circuit symbol be active before the output of that symbol is active—also, Boolean product or intersection.

AND function   The function that derives from the definition of AND.

AND gate   A physical device that performs the electrical equivalent of the AND function.

AND laws   A set of Boolean identities based on the AND function.

Antiphase   A term used in this book to mean complemented triggering of a device relative to a reference system as, for example, an FET or RET D-latch.

A-OPS   Asynchronous one-hot programmable sequencer.

Arbiter or arbiter module   A device that is designed to control access to a protected system by arbitration of contending signals.

Array algebra   The algebra of Boolean arrays and matrices associated with the design of STT machines.

Array logic   Any of a variety of logic devices, such as ROMs, PLAs, or PALs, that are composed of an AND array and an OR array (see **Programmable logic device** or **PLD**).

ASIC   Application specific IC.

Assert   Activate.

Assertion level   Activation level.

Associative law   A law of Boolean algebra that states that the operational sequence as indicated by the location of parentheses in a p-term or s-term does not matter.

Associative pattern   An XOR pattern in a K-map that allows a term or variable in an XOR or EQV function to be looped out (associated) with the same term or variable in an adjacent cell provided that the XOR or EQV connective is preserved in the process.

Asynchronous   Clock independent or self- timed—having no fixed time relationship.

Asynchronous input   An input that can change at any time, particularly during the sampling interval of the enabling input.

Asynchronous override   An input such as preset or clear that, when activated, interrupts the normal operation of a memory element.

Asynchronous parallel load   The parallel loading of a register or counter by means of the asynchronous PR and CL overrides of the flip-flops.

Basic cell   A basic memory cell, composed of either cross-coupled NAND gates or cross-coupled NOR gates, used in the design of other asynchronous FSMs including flip-flops.

Binary   A number system of radix two; having two values or states.

Binary code   A combination of bits that represent alphanumeric and arithmetic information.

Binary coded decimal (BCD)   A four-bit, 10-word decimal code that is weighted 8, 4, 2, 1 and is used to represent decimal digits as binary numbers.

Binary coded hexadecimal (BCH)   The hexadecimal number system used to represent bit patterns in binary.

Binary coded octal (BCO)   The octal number system used to represent bit patterns in binary.

Binary word   A linear array of juxtaposed bits that represents a number or that conveys an item of information.

Bit   A binary digit.

Bit slice   Partitioned into identical parts such that each part operates on 1 bit in a multibit word—part of a cascaded system of identical parts.

Boolean algebra   The mathematics of logic attributed to the mathematician George Boole (1815–1864).

Boolean product   AND or intersection operation.

Boolean sum   OR or union operation.

BOOZER   BOOlean ZEro-one Reduction—a multioutput logic minimizer that accepts entered variables or canonical data.

Boundary   The separation of logic domains in a K-map.

Bounded pulse   A pulse with both lower and upper limits to its width.

Branching condition (BC)   The input requirements that control a state-to-state transition in an FSM.

Branching path   A state-to-state transition path in a state diagram.

Buffer   A line driver.

Buffer state   A state (in a state diagram) whose only purpose is to remove a race condition.

Burst-mode FSM   An FSM whose branching paths in a state diagram are labeled with prescribed input and output signal transitions as opposed to their logic values as in the classical notation.

Bus arbiter   A two-input/two output arbiter that grants on request an output on a first-in/first-out basis but will grant access to a second request only after the first request is withdrawn.

Byte   A group of 8 bits.

Call module   A module designed to control access to a protected system by issuing a request for access to the system and then granting access after receiving acknowledgment of that request.

Canonical   Made up of terms that are either all minterms or all maxterms.

Canonical truth table   A 1's and 0's truth table consisting exclusively of minterms or maxterms.

CAPS   Cascadable asynchronous programmable sequencers.

Cardinality   The number of prime implements (p-term or s-term cover) representing a function.

Cascade   To combine identical devices in series such that any one device drives another; to bit-slice.

C-element   A two-input rendezvous module (RMOD) that operates outside of the fundamental mode such that its inputs are permitted to change simultaneously.

Cell   The intersection of all possible domains of a K-map.

Circuit   A combination of elements (e.g., logic devices) that are connected together to perform a specific operation.

CL or CLR   Clear.

CLEAR   An asynchronous input that, when activated, forces the internal state of the device to logic 0.

Clock (CK)   A regular source of pulses that control the timing operations of a synchronous sequential machine.

Clock skew   A phenomenon that is generally associated with high-frequency clock distribution problems in synchronous sequential systems.

CMOS   Complementary configured MOSFET in which both NMOS and PMOS are used.

CNT   Mnemonic for count.

Code converter   A device designed to convert one binary code to another.

Collapsed truth table   A truth table containing irrelevant input symbols.

Combinational hazard   A hazard that is produced within a combinational logic circuit.

Combinational logic   A configuration of logic devices in which the outputs occur in direct, immediate response to the inputs without feedback.

Commutative law   The Boolean law that states that the order in which variables are represented in a p-term or s-term does not matter.

Compatibility   A condition where the input to a logic device and the input requirement of the device are of the same activation level, that is, are in logic agreement.

Complementary C-element   A Muller C-element in which one of its inputs required and active low input.

Complementary metal oxide semiconductor (CMOS)   A form of MOS that uses both p- and n-channel transistors (in pairs) to form logic gates.

Complementation   A condition that results from logic incompatibility; the mixed-logic equivalent of the NOT operation.

Composite output map   A K-map that contains entries representing multiple outputs.

Computer   A digital device that can be programmed to perform a variety of tasks (e.g., computations) at high speed.

Concatenation   Act of linking together or being linked together in a series.

Conditional branching   State-to-state transitions that depend on the input status of the FSM.

Conditional output   An output that depends on one or more external inputs.

Conjugate gate forms   A pair of logic circuit symbols that derive from the same physical gate and that satisfy the DeMorgan relations.

Connective   A Boolean operator symbol (e.g., +, ·, ⊕, etc.).

Consensus law   A law in Boolean algebra that allows simplification by removal of a redundant term.

Consensus term   The redundant term that appears in a function obeying the consensus law.

Controlled inverter   An XOR gate that is used in either the inverter or transfer mode.

Controller   That part of a digital system that controls the data path devices.

Conventional K-map   A K-map whose cell entries are exclusively 1's and 0's.

Counteracting delay   A delay placed on an external feedback path to eliminate an E-hazard or d-trio.

Counter   A combinational logic device whose function it is to count in binary some count sequence.

Coupled term   One of two terms containing only one coupled variable.

Coupled variable   A variable that appears complemented in one term of an expression (SOP or POS) and that also appears uncomplemented in another term of the same expression.

Cover   A set of terms that covers all minterms or maxterms of a function.

CPLD   Complex PLD.

Critical race   A race condition in an asynchronous FSM that can result in a transition to an erroneous state.

Cross branching   Multiple transition paths from one or more states in the state diagram (or state table) of a sequential machine whereby unit distance coding of states is not possible.

Cube   A p-term containing a set of 2^n minterms in reduced or minimized form. Also an s-term containing a set of 2^n maxterms in reduced or minimized form.

Cycle   Two or more successive and uninterrupted state-to-state transitions in an asynchronous sequential machine, usually not controlled by input conditions.

Cycle state   A state among others in state-to-state transitions not controlled by external inputs.

Data bus   A parallel set of conductors that are capable of transmitting or receiving data between two parts of a system.

Data path   The part of a digital system that is controlled by a controller.

Data path unit   The group of logic devices that comprise the data path.

Data selector   A multiplexer.

Data-triggered   A term that refers to flip-flops triggered by external inputs (no clock) as in the pulse mode.

DCL   Digital combination lock.

Deactivate   To make inactive.

Debounce   To remove the noise that is produced by a mechanical switch.

Debouncing circuit   A circuit that is used to debounce a mechanical switch.

Decade   A quantity of 10.

Decoder   A combinational logic device that will activate a particular minterm code output line determined by the binary code input. A demultiplexer.

Delay   The time elapsing between related events in progress.

Delay circuit   A circuit whose purpose it is to delay a signal for a specified period.

DeMorgan relations   Mixed logic expressions of DeMorgan's laws.

DeMorgan's laws   A property that states that the complement of the Boolean product of terms is equal to the Boolean sum of their complements; or that states that the complement of the Boolean sum of terms is the Boolean product of their complements.

Demultiplexer   A decoder.

D flip-flop   A 1-bit memory device whose output value is set to the D input value on the triggering edge of the clock signal.

DFLOP module   A memory element used in an EAIC system that has characteristics similar to that of a D flip-flop but that delivers a "data ready" signal.

Diagonal pattern   An XOR pattern formed by identical EV subfunctions in any two diagonally located cells of a K-map whose coordinates differ by 2 bits.

Digit   A single symbol in a number system.

Digital   Related to discrete quantities.

Digital combination lock   A sequence recognizer that can be used to unlock or lock something.

Digital engineering design   The design and analysis of digital devices.

Digital signal   A logic waveform composed of discrete logic levels (e.g., a binary digital signal).

Diode   A two-terminal passive device consisting of a p–n junction that permits significant current to flow only in one direction.

Disjoint   As used in "mutually disjoint" to mean a set of p-terms whose ANDed values taken two at a time are always logic zero. Thus, mutually disjoint terms never take logic 1 at the same time.

Distributive law   The dual of the factoring law where both are categorized as a distributive law.

DMUX   Demultiplexer (see **Decoder**).

Domain   A range of logic influence or control.

Domain boundary   The vertical or horizontal line or edge of a K-map.

Don't care   A nonessential minterm or maxterm, denoted by the symbol ϕ, which can take either a logic 1 or logic 0 value. Also a delimiter that, when attached to a variable or term, renders that variable or term nonessential to the parent function.

Don't care state   A state having only output transition paths and that is not part of the primary (essential) state sequence in the state diagram representing an asynchronous FSM.

Driver   A one-input device whose output can drive substantially more inputs than a standard gate. A buffer.

D-trio   A type of essential hazard that causes a fundamental mode machine to transit to the correct state via an unauthorized path.

Duality   A property of Boolean algebra that results when the AND and OR operators (or XOR and EQV operators) are interchanged simultaneously with the interchange of 1's and 0's.

Dual relations   Two Boolean expressions that can be derived one from the other by duality.

Duty cycle   In a periodic waveform, the percentage of time the waveform is active.

Dyad   A grouping of two logically adjacent minterms or maxterms.

Dynamic hazard   Multiple glitches in the output such that the output logic levels are different before and after an input change and which are usually produced in multilevel circuits involving three or more asymmetric paths (delay-wise) of the input to the output.

EAIC system   Externally asynchronous/internally clocked system.

Edge-triggered flip-flop   A FLIP-FLOP that is triggered on either the rising edge or falling edge of the clock waveform and that exhibits the data lockout feature.

E-hazard   Essential hazard.

EN   Enable.

Enable   An input that is used to enable (or disable) a logic device, or that permits the device to operate normally.

Endless cycle   An oscillation that occurs between states in an asynchronous FSM.

Entered variable (EV)   A variable entered into a K-map.

EPI   Essential prime implicant.

EPLD   Erasable PLD.

EPROM   Erasable programmable read-only memory.

Equivalence   The output of a two-input logic gate that is active if, and only if, its inputs are logically equivalent (i.e., both active or both inactive).

EQV   Equivalence.

EQV function   The function that derives from the definition of equivalence and given the symbol $\odot$ or $\overline{\oplus}$ as used in this text.

EQV gate   A physical device that performs the electrical equivalent of the EQV function.

EQV laws   A set of Boolean identities based on the EQV function.

Erasable programmable read-only memory (EPROM)   A ROM that can be programmed many times.

ESPRESSO logic minimizer   A software minimization tool that supports advanced heuristic algorithms of multioutput Boolean functions but does not accept entered variables.

Essential hazard (E-hazard)   A disruptive sequential hazard that can occur as a result of an unintended, explicitly located delay in an asynchronous FSM that has at least three states.

Essential prime implicant (EPI)   A PI that must be used to achieve minimum cover.

EV   Entered variable.

EV K-map   A K-map that contains EVs.

EV truth table   A truth table containing EVs.

Even parity   An even number of 1's (or 0's) in a binary word depending on how even parity is defined.

EVM Entered variable K-map.

Excitation table A state transition table relating the branching paths to the branching condition values given in the state diagram for a memory element.

Exclusive OR (XOR) A two-variable function that is active iff one of the two variables is active.

EXL-Sim A powerful, fully featured, interactive, schematic capture, and simulation software ideally suited for use with this text (see link exlsim.com).

Extended burst-mode FSM Use of directed don't care branching conditions that allow the designer to specify that an input change may or may not happen in a given input burst.

Factoring law The Boolean law that permits a variable to be factored out of two or more p-terms that contain that variable in an SOP or XOR expression.

Falling edge-triggered (FET) Activation of a device on the falling edge of the triggering (sampling) variable.

Fan-in The maximum number of inputs a gate may have.

Fan-out The maximum number of equivalent gate inputs that a logic gate output can drive.

Feedback path A signal path of a PS variable from the memory output to the NS input.

FET Falling edge-triggered. Also, field effect transistor.

FF Flip-flop.

Field programmable gate array (FPGA) A complex PLD that may contain a variety of primitive devices such as discrete gates, MUXs, and flip-flops.

Field programmable logic array (FPLA) One-time user programmable PLA.

Finite state machine (FSM) A sequential machine that has a finite number of states into which it can reside.

Flip-flop (FF) A one-bit memory element that exhibits sequential behavior exclusively controlled by a clock input.

Flow table A tabular realization of a state diagram representing the sequential nature of an FSM.

Fly state A state (in a state diagram) whose only purpose is to remove a race condition. A buffer state.

Forward bias A voltage applied to a p–n junction diode in a direction as to cause the diode to conduct (turn ON).

FPGA Field programmable gate array.

FPLA Field programmable logic array.

FPLS Field programmable logic sequencer.

Frequency, f The number of waveform cycles per unit time in Hz.

FSM Finite state machine, either synchronous or asynchronous.

Fully documented state diagram A state diagram that specifies all input branching conditions and output conditions in literal or mnemonic form, that satisfies the sum rule and mutually exclusive requirement, and that has been given a proper state code assignment.

Function　A Boolean expression representing specific binary operations.

Functional partition　A diagram that gives the division of device responsibility in a digital system.

Function hazard　A hazard that is produced when two or more coupled variables change in close proximity to each other—an output response to competing coupled input changes.

Fundamental mode　The operational condition of an asynchronous FSM in which no input change is permitted to occur until the FSM has stabilized following any previous input change.

GAL　Generic array logic.

Gate　A physical device (circuit) that performs the electrical equivalent of a logic function. Also, one of three terminals of a MOSFET.

Gate/input tally　The gate and input count associated with a given logic expression—the gate tally may or may not include inverters, but the input count must include both external and internal inputs.

Gate-minimum logic　Logic requiring a minimum number of gates, and may include XOR and EQV gates in addition to two-level logic.

Gate path delay　The interval of time required for the output of a gate to respond to an input signal change.

Glitch　An unwanted transient in an otherwise steady-state signal.

Go/no-go configuration　A single input controlling the hold and exit conditions of a state in a state diagram.

Gray code　A reflective unit distance code.

Ground　A reference voltage level usually taken to be zero volts.

Hamming distance　As used in this text, the number of state variables required to change during a given state-to-state transition in an FSM.

Handshake arbiter　A versatile arbiter that receives a done (acknowledgment) signal and issues a grant signal following the successful receipt of a request from a protected system.

Handshake interface　A configuration between two devices whereby the outputs of one device are the inputs to the other and vice versa.

Hang state　An isolated state into which an FSM can reside stably but which is not part of the authorized routine.

Hardware description language (HDL)　A high-level programming language with specialized structures for modeling hardware.

Hazard　A glitch or unauthorized transition that is caused by an asymmetric path delay via an inverter, gate, or lead during a logic operation.

Hazard cover　The redundant cover (consensus term) that removes a static hazard.

HDL　Hardware description language.

Heuristic　By empirical means or by discovery.

Hold condition   Branching from a given state back into itself or the input requirements necessary to effect such branching action.

Hold time   The interval of time immediately following the transition point during which the data inputs must remain logically stable to ensure that the intended transition of the FSM will be successfully completed.

Huffman circuit   An asynchronous FSM that operates in the fundamental mode.

HV   High voltage.

Hybrid function   Any function containing both SOP and POS terms.

IC   Integrated circuit.

Implicant   A term in a reduced or minimized expression.

Inactive   Not active and implying logic 0.

Inactive state   The logic 0 state of a logic device.

Inactive transition point   The point in a logic waveform where a digital device passes from the active state to the inactive state.

Incompatibility   A condition where the input to a logic device and the input requirement of that device are of opposite activation levels.

Incompletely specified function   A function that contains nonessential minterms or maxterms (see **Don't care**).

Indirect path   The path taken by the initiator input to the race gate (RG) via the second y-variable invariant in the development of an E-hazard or d-trio.

Inertial delay circuit   A delay circuit based mainly on R–C components.

Initialize   To drive a logic circuit into a beginning or reference state.

Initiator input   The external input whose single change initiates an E-hazard or d-trio.

Input   A signal or line into a logic device that controls the operation of that device.

Intersection   AND operation.

Inversion   The inverting of a signal from HV to LV or vice versa.

Inverter   A physical device that performs voltage inversion in the physical domain or that performs logic level conversion in the logic domain.

I/O   Input/output.

IPG   Indirect path gate.

Irredundant   Not redundant, as applied to an absolute minimum Boolean expression.

Irrelevant input   An input whose presence in a function is nonessential.

Island   A K-map entry that must be looped out of a single cell.

Iterative   Repeated many times to achieve a specific goal.

Juxtapose   To place side by side.

Karnaugh map (K-map)   Graphical representation of a logic function named after M. Karnaugh (1953).

K-map   Karnaugh map.

Latch   A name given to certain types of memory elements as, for example, the D latch.

Latency   The time required to complete an operation in a sequential machine.

LD   Mnemonic for load.

Least significant bit (LSB)   The bit (usually at the extreme right) of a binary word that has the lowest positional weight.

Level   A term used when specifying the number of gate path delays of a logic function (from input to output) usually exclusive of inverters. See, for example, two-level logic.

Level triggered   Rising edge triggered (RET) or falling edge triggered (FET).

Linear state machine   An FSM with a linear array of states.

Line driver   A device used to boost and sharpen a signal so as to avoid fan-out problems.

Logic   The functional capability of a digital device that is interpreted as either a logic 1 or logic 0.

Logic adjacency   Two logic states whose state variables differ from each other by only 1 bit.

Logic circuit   A digital circuit that performs the electrical equivalent of some logic function or process.

Logic diagram   A digital circuit schematic consisting of an interconnection of logic symbols.

Logic instability   The inability of a logic circuit to maintain a stable logic condition. Also, an oscillatory condition in an asynchronous FSM.

Logic level   Logic status indicating either positive logic or negative logic.

Logic level conversion   The act of converting from positive logic to negative logic or vice versa.

Logic map   Any of a variety of graphical representations of a logic function.

Logic noise   Undesirable signal fluctuations produced within a logic circuit following input changes.

Logic state   A unique set of binary values that characterize the logic status of a machine at some point in time.

Logic waveform   A rectangular waveform between active and inactive logic states.

Loop-out   The action that identifies a prime implicant in a K-map.

Loop-out protocol   A minimization procedure whereby the largest group of logically adjacent minterms or maxterms are looped out in the order of increasing n ($n = 0, 1, 2, 3, \ldots$).

LPD   Lumped path delay.

LSB   Least significant bit.

LSD   Least significant digit.

LSI   Large-scale integration.

Lumped path delay (LPD) model   A model, which is applicable to FSMs that operate in the fundamental mode, characterized by a lumped memory element for each state variable/feedback path.

LV   Low voltage.

MAC module Microprogrammable asynchronous controller module.

Majority function A function that becomes active when a majority of its variables become active.

Majority gate A logic gate that yields a majority function.

Map Usually a Karnaugh map.

Map compression A reduction in the order of a K-map usually involving EVs.

Mapping algorithm In FSM design, the procedure to obtain the NS functions by ANDing the memory input logic value in the excitation table with the corresponding branching condition in the state diagram for the FSM to be designed, and entering the result in the appropriate cell of the NS K-map.

Maxterm A POS term that contains all the variables of the function.

Maxterm code A code in which complemented variables are assigned logic 1 and uncomplemented variables are assigned logic 0—the opposite of minterm code.

MD/ME stage Metastable detection/mutually exclusive stage.

MDS. Metastable detection stage.

Mealy machine An FSM that conforms to the Mealy model.

Mealy model The general model for a sequential machine where the output state depends on the input state as well as the present state.

Mealy output A conditional output.

Memory The ability of a digital device to store and retrieve binary words on command.

Memory element A device for storing and retrieving 1 bit of information on command in a synchronous or asynchronous FSM. Also in asynchronous FSM terminology only, a fictitious lumped path delay, a basic cell or C-element.

Merge The concatenation of buses to form a larger bus.

Merging of states In a state diagram, the act of combining states to produce fewer states.

Metal-oxide-semiconductor The material constitution of an important logic family (MOS) used in IC construction.

Metastability An unresolved state of an FSM which resides between a set and a reset condition or which becomes logically unstable for an unpredictable period of time, in either case.

Metastable exit time The unpredictable time interval between entrance into and exit from the metastable state.

MEV Map entered variable.

Minimization The process of reducing a logic function to its simplest form.

Minimum cover The optimally reduced representation of a logic expression.

Minterm A term in an SOP expression where all variables of the expression are represented in either complemented or uncomplemented form.

Minterm code A logic code in which complemented variables are assigned logic 0 whereas uncomplemented variables are assigned logic 1—the opposite of maxterm code.

Mixed logic   The combined use of the positive and negative logic systems.

Mixed-rail output   Dual, logically equal outputs of a device (e.g., a basic cell) where one output is issued active high whereas the other is issued active low, but which are not issued simultaneously.

Mnemonic   A short single group of symbols (usually letters) used to convey a meaning.

Mnemonic state diagram   A fully documented state diagram.

Model   The means by which the major components and their interconnections are represented for a digital machine or system.

Module   A device that performs a specific function and that can be added to or removed from a system to alter the system's capability. A common example is an arbiter module.

Monad   A minterm (or maxterm) that is not logically adjacent to any other minterm (or maxterm) in a K-map.

Moore machine   A sequential machine that conforms to the Moore model.

Moore model   A degenerate form of the Mealy (general) model in which the output state depends only on the present state and which is independent of external inputs.

Moore output   An unconditional output.

MOS   Metal-oxide semiconductor.

MOSFET   Metal-oxide semiconductor field effect transistor.

Most significant bit (MSB)   The extreme left bit of a binary word that has the highest positional weight.

MSB   Most significant bit.

MSD   Most significant digit.

MSI   Medium-scale integration.

MTBF   Mean time between failures.

Muller C-element   A rendezvous module (RMOD) that operates outside of the fundamental mode and that changes activation levels only after all inputs change to the same activation level.

Muller circuit   A speed-independent circuit that operates outside of the fundamental mode.

Multilevel logic minimization   Minimization involving more than two levels of path delay.

Multiple-output minimization   Optimization of more than one output expression from the same logic device.

Multiplex   To select or gate (on a time-shared basis) data from two or more sources onto a single line or transmission path.

Multiplexer   A device that multiplexes data.

Mutually exclusive requirement   A requirement in state diagram construction that forbids overlapping branching conditions (BCs)—i.e., it forbids the use of BCs shared between two or more branching paths.

MUX   Multiplexer.

NAND-centered basic cell   Cross-coupled NAND gates forming a basic cell.

NAND gate   A physical device that performs the electrical equivalent of the NOT AND function.

NAND/INV logic   Combinational logic consisting exclusively of NAND gates and inverters.

Negative logic   A logic system in which high voltage (HV) corresponds to logic 0 and low voltage (LV) corresponds to logic 1. The opposite of positive logic.

Negative pulse   A 1-0-1 pulse.

Nested cell   A basic cell used as the memory in an asynchronous FSM design.

Nested element model   A model in which memory elements are embedded into an FSM.

Nested machine   Any asynchronous machine that serves as the memory in the design of a larger sequential machine. Any FSM that is embedded within another.

Next state (NS)   A state that follows the present state (PS) in a sequence of states.

Next state forming logic   The logic hardware in a sequential machine whose purpose it is to generate the next state function input to the memory.

Next state function   The logic function that defines the next state of an FSM given the present state.

Next state map   A composite K-map where the entries for each cell are the next state subfunction for the present state represented by the coordinates of that cell.

Next state variable   The variable representing the next state function.

NMOS   An n-channel MOSFET.

Noise immunity   The ability of a logic circuit to reject unwanted signals.

Noise margin   The maximum voltage fluctuation that can be tolerated in a digital signal without crossing the switching threshold of the switching device.

NOR-centered basic cell   Cross-coupled NOR gates forming a basic cell.

NOR gate   A physical device that performs the electrical equivalent of the NOT OR function.

NOR/INV logic   Combinational logic consisting exclusively of NOR gates and inverters.

NOT function   An operation that is the logic equivalent of complementation.

NOT laws   A set of Boolean identities based on the NOT function.

NS   Next state.

Octad   A grouping of eight logically adjacent minterms or maxterms.

Odd parity   An odd number of 1's or 0's depending on how odd parity is defined.

Offset pattern   An XOR pattern in a K-map in which identical subfunctions are located in two nondiagonal cells that differ in cell coordinates by 2 bits.

O-HAPS.   One-hot asynchronous programmable sequencer.

One-hot code   A nonweighted code in which there exists only one "1" in each word of the code.

One-hot design method   The use of the one-hot code for synchronous and asynchronous FSM design.

One-hot-plus-zero   One-hot code plus the all-zero state.

Operand   A number or quantity that is to be operated on.

Operation table   A table that defines the functionality of a device.

Operator   A Boolean connective.

OPI   Optional prime implicant.

Optional prime implicant (OPI)   A PI whose presence in a minimum function produces alternative minimum cover.

OR   An operator requiring that the output of an OR gate be active if one or more of its inputs are active.

Order   Refers to the number of variables on the axes of a K-map.

OR function   A function that derives from the definition of OR.

ORG   Output race glitch.

OR gate   A physical device that performs the electrical equivalent of the OR function.

OR laws   A set of Boolean identities based on the OR function.

Outbranching   Branching from a state exclusive of the HOLD branching condition.

Output   A concluding signal issued by a digital device.

Output forming logic   The logic hardware in a sequential machine whose purpose it is to generate the output signals.

Output race glitch (ORG)   An internally initiated function hazard that is produced by a race condition in a sequential machine.

Packing density   The practical limit to which switches of the same logic family can be packed in an IC chip.

PAL   Programmable array logic (registered trademark of Advanced Micro Devices, Inc.).

Parity   Related to the existence of an even or odd number of 1's or 0's in a binary word.

Parity bit   A bit appended to a binary word to detect, create, or remove even or odd parity.

Pass transistor switch   An MOS transistor switch that functions as a nonrestoring switching device and that does not invert a voltage signal. A transmission gate.

PDP   Power-delay product.

Period   The time in seconds [s] between repeating portions of a waveform, hence, the inverse of the frequency.

Physical truth table   An I/O specification table based on a physically measurable quantity such as voltage.

PI   Prime implicant.

Pipeline   A processing scheme where each task is allocated to specific hardware (joined in a line) and to a specific time slot.

PISO   Parallel-in/serial-out operation mode of a register.

PLA   Programmable logic array.

Planar format   A two-dimensional K-map array used to minimize functions of more than four variables.

PLD   Programmable logic device such as PALs, FPLAs, FPGAs, ROMs, EPROMs, and other CPLDs.

PLS   Programmable logic sequencer.

PMOS   A p-channel MOSFET.

p–n junction diode   (see **Diode**).

Polarized mnemonic   A contracted signal name onto which is attached an activation level indicator.

Port   An entry or exit element to an entity (e.g., the name given to an input signal in a VHDL declaration).

POS   Product-of-sums.

POS HAZARD   A static-0 hazard.

Positive logic   Logic system in which HV corresponds to logic 1 and LV corresponds to logic 0.

Positive pulse   A 0-1-0 pulse.

Power, P   The product of voltage, V, and current, I, given in units of watts [W].

Power-delay product (PDP)   The average power dissipated by a logic device multiplied by its propagation delay time.

PR or PRE   PRESET.

Present state (PS)   The logic state into which the FSM resides at a given instant.

Present-state/next-state (PS/NS) table   A table that is produced from the next state K-maps and that is used to construct a fully documented state diagram in an FSM analysis.

Preset   An asynchronous input that is used in to set an FSM to a logic 1 condition.

Prime implicant (PI)   A cube that cannot be combined with any other cube in any way to produce a cube of fewer variables.

Primitive   A discrete logic device such as a gate, MUX, decoder, etc.

Priority stand-alone arbiter   An arbiter design for a fixed number of inputs and that cannot be combined with other such arbiters to increase that number of inputs.

Product-of-sums (POS)   The ANDing of ORed terms in a Boolean expression.

Programmable logic array (PLA)   Any PLD that can be programmed in both the AND and OR planes.

Programmable logic device (PLD)   Any two-level combinational array logic device from the families of ROMs, PLAs, PALs or FPGAs, etc.

Programmable read-only memory (PROM)   A once-only user-programmable ROM.

PROM   Programmable read-only memory.

Propagation delay   In a logic device (e.g., a gate), the time interval of an output response to an input signal.

PS   Present state.

PS/NS   Present-state/next-state.

P-term   A Boolean product term—one consisting only of ANDed literals.

P-term table   A table that consists of p-terms, inputs and outputs and that is used to program PLA type devices.

Pull-down resistor   A resistor that causes a signal on a line to remain at low voltage.

Pull-up resistor   A resistor that causes a signal on a line to remain at high voltage.

Pulse   An abrupt change from one level to another followed by an opposite abrupt change.

Pulse mode   An operational condition for an asynchronous FSM where the inputs are required to be discrete nonoverlapping pulse signals.

Pulse width   The active duration of a positive pulse or the inactive duration of a negative pulse.

Quad   A grouping of four logically adjacent minterms or maxterms.

R   Request or data ready signal. Also reset or resistance.

Race condition   A condition in an asynchronous sequential circuit where the transition from one state to another involves two or more alternative paths.

Race gate   The gate to which two or more input signals are in race contention.

Race path   Any path that can be taken in a race condition.

Race state   Any state through which an FSM may transit during a race condition.

RAM   Random access memory.

Random access memory (RAM)   A read/write memory system in which all memory locations can be accessed directly independent of other memory locations.

R-C   Resistance/capacitance or resistor/capacitor.

Read-only memory (ROM)   A PLD that can be mask programmed only in the OR plane.

Redundant cover   Nonessential and nonoptional cover in a function representation.

Redundant prime implicant   A PI that yields redundant cover.

Rendezvous module (RMOD)   An asynchronous state machine whose output becomes active when all external inputs become active and becomes inactive when all external inputs become inactive.

Reset   A logic 0 condition or an input to a logic device that sets it to a logic 0 condition.

Residue   The part of term that remains when the coupled variable is removed (see **Consensus term**).

Resistance, R. The voltage drop across a conducting element divided by current through the element (in ohms).

RET   Rising edge triggered.

Reverse bias   A voltage applied to a p–n junction diode in a direction as to minimize conduction across the junction.

Reverse saturation current The current through a p–n junction diode under reverse bias.

RG (See **Race gate**)

Rise time The length of time it takes a voltage (or current) signal to change from 10% to 90% of its high value.

Rising edge triggered (RET) Activation of a logic device on the rising edge of the triggering variable.

RMOD Rendezvous module.

ROM Read-only memory.

Rotating token arbiter A multistage arbiter in which each stage sampled by a rotating token signal results in the issuance of a grant from a given stage provided a request input to that stage is active at the time the token enters that stage.

RPI Redundant prime implicant.

Runt pulse Any pulse that falls short of reaching the switching threshold of a device into which it is introduced.

S SET.

SAM (See **State array machine**)

Sampling interval Sum of the setup and hold times.

Sampling variable The last variable to change in initiating a state-to-state transition in an FSM.

Sanity circuit A circuit that is used to initialize an FSM into a particular state; usually a resistor/capacitor (R/C) type circuit.

Schmitt trigger An electronic gate with hysteresis and high noise immunity that is used to "square up" pulses.

Selector module A device whose function it is to steer one of two input signals to either one of two outputs depending on whether a specific input is active or inactive.

Sequence detector (recognizer) A sequential machine that is designed to recognize a particular sequence of input signals.

Sequential hazard An essential hazard.

Sequential machine Any digital machine with feedback paths whose operation is a function of both its history and present input data.

SET A logic 1 condition or an input to a logic device that sets it to a logic 1 condition.

Set-up time The interval of time prior to the transition point during which all data inputs must remain stable at their proper logic level to ensure that the intended transition will be initiated.

S-hazard A static hazard.

Shared prime implicant (SPI) Two identical PIs appearing in two or more NS and/or output functions from the same FSM.

Single transition time (STT) A state-to-state transition in an asynchronous FSM that occurs in the shortest possible time, that is without passing through a race state.

SOP   Sum-of-products.

SOP hazard   Static-1 hazard.

Source   The origin of a digital signal.

SPDT switch   Single-pole/double-throw switch.

Square wave   A rectangular waveform.

SRAM   Static RAM.

Stability criteria   The requirements that determine if an asynchronous FSM, operated in the fundamental mode, is stable in a given state.

Stable state   Any logic state of an asynchronous FSM that satisfies the stability criteria.

Stack format   A three-dimensional array of conventional fourth-order K-maps used for function minimization of more than four variables.

State   A unique set of binary values that characterize the logic status of a machine at some point in time.

State adjacency set   Any 2^n set of logically adjacent states of an FSM.

State array machine (SAM)   A 2^n array of states such that each state in the array has transition paths only to states that are logically adjacent.

State code assignment   Unique set of code words that are assigned to an FSM to characterize its logic status.

State diagram   The diagram or chart of an FSM that shows the state sequence, branching conditions, and output information necessary to describe its sequential behavior.

State machine   A finite state machine (FSM). A sequential machine.

State identifier   Any symbol (e.g., numerical or alphabetical) that is used to represent or identify a state in a state diagram.

State table   Tabular representation of a state diagram.

State transition table   (see **Excitation table**).

State variable   Any variable whose logic value contributes to the logic status of a machine at any point in time. Any bit in the state code assignment of a state diagram.

Static hazard   An unwanted glitch in an otherwise steady-state signal that is produced by an input change propagating along asymmetric path delays through inverters or gates.

Static-1 hazard   A glitch that occurs in an otherwise steady-state 1 output signal from SOP logic due to a change in an input for which there are two asymmetric paths (delay-wise) to the output.

Static-0 hazard   A glitch that occurs in an otherwise steady-state 0 output signal from POS logic due to a change in an input for which there are two asymmetric paths (delay-wise) to the output.

Static RAM   A nonvolatile form of RAM—does not need periodic refreshing to hold its information.

Steering logic   Logic based primarily on transmission gate switches.

S-term   A Boolean sum term—one containing only ORed literals.

STT   Single transition time.

Sum-of-products (SOP)   The ORing of ANDed terms in a Boolean expression.

Sum rule   A rule in state diagram construction that requires that all possible branching conditions be accounted for.

Switching speed   A device parameter that is related to its propagation delay time.

Synchronizer circuit   A logic circuit (usually a D flip-flop) that is used to synchronize an input with respect to a clock signal.

Synchronous machine   A sequential machine that is clock driven.

System level design   A design that includes controller and data path sections.

Tabular minimization   A minimization procedure that uses tables exclusively.

TCM   (See **Timing control machine**)

T flip-flop   A flip-flop that operates in either the toggle or hold mode.

Throughput   The time required to produce an output response due to an input change.

Time constant   The product of resistance and capacitance given in units of seconds [s]—a measure of the recovery time of an R–C circuit.

Timing control machine (TCM)   A three-state resolver FSM whose function it is to control the SAM transitions via a handshake configuration in the MAC module.

Timing diagram   A set of logic waveforms showing the time relationships between two or more logic signals.

Toggle   Repeated but controlled transitions between any two states, as between the set and reset states.

Toggle module   A flip-flop that is configured to toggle only. Also, a divide-by-2 counter.

Trans-HI module   A transparent high (RET) D latch.

Trans-LO module   A transparent low (FET) D latch.

Transistor   A three-terminal switching device that exhibits current or voltage gain.

Transition   In a digital machine, a change of one state (or level) to another.

Transmission gate   A pass transistor switch.

Transparent D latch   A two-state D flip-flop in which the output, Q, tracks the input, D, when clock is active if RET or when clock is inactive if FET.

Tree   Combining of like gates or C-elements to accommodate multiple inputs or to overcome fan-in limitations.

Triggering threshold   The point beyond which a transition takes place.

Triggering variable   Sampling (enabling) variable.

Tri-state bus   The wire-ORed output lines from a multiplexed scheme of PLDs having tri-state enables.

Tri-state driver   An active logic device that operates in either a disconnect mode or an inverting (or noninverting) mode.

Truth table   A table that provides an output value for each possible input condition to a combinational logic device.

Two-level logic   Logic consisting of only one ANDing and one ORing stage.

Unconditional branching   State-to-state transitions that take place independent of the input status of the FSM.

Unconditional output   An output of an FSM that does not depend on an input signal. A Moore output.

Union   OR operation.

Unit distance code   A code in which each state in the code is surrounded by logically adjacent states.

Unstable state   Any logic state in an asynchronous FSM that does not satisfy the stability criteria.

VEM   Variable entered map.

Very-large-scale integrated circuits   IC chips that contain thousands to millions of gates.

VHDL   VHSIC hardware description language.

VHSIC   Very-high-speed integrated circuit.

VLSI   Very-large-scale integrated circuit.

Voltage, *V*   The potential difference between two points, in units of volts [V]. Also, the work required to move a positive test charge against an electric field.

Voltage waveform   A voltage waveform in which rise and fall times exist.

Wireless connection feature   In schematic capture, the simplifying feature that permits input and output labels to be presented separately from corresponding labels used throughout a circuit—thus, no wire connections are needed to convey connectivity.

WSI circuits   Wafer-scale integrated circuits.

XNOR   Exclusive NOR (see **Equivalence** and **EQV**).

XOR   Exclusive OR.

XOR function   The function that derives from the definition of Exclusive OR.

XOR gate   A physical device that performs the electrical equivalent of the XOR function.

XOR laws   A set of Boolean identities that are based on the XOR function.

XOR pattern   Any of four possible K-map patterns that result in XOR type functions.

Author Biography

Professor Richard Tinder's teaching interests have been highly varied over his tenure at Washington State University (WSU). They have included crystallography, thermodynamics of solids (both equilibrium and irreversible thermodynamics), solid state electronics, tensor properties of crystals, dislocation theory, solid state direct energy conversion (mainly solar cell theory, thermoelectric effects, and fuel cells), general materials science, electromagnetics, and analog and digital circuit theory. For more than 20 years, he taught logic design at the entry, intermediate, and advanced levels and has published a major text in the area titled *Engineering Digital Design, 2nd Ed. Revised*. He has conducted research and published in the areas of tensor properties of solids, surface physics, shock dynamics of solids, milli–micro plastic flow in metallic single crystals, high-speed asynchronous (clock-independent) state machine design, and Boolean algebra (specifically XOR algebra and graphics). Recently, he has published two books: *Relativistic Flight Mechanics and Space Travel* and *Tensor Properties of Solids—Phenomenological Development of the Tensor Properties of Crystals*.

Professor Tinder received his bachelor's, master's, and doctoral degrees from the University of California, Berkeley. In the early 1970s, he spent one year as a visiting faculty member at the University of California, Davis, in what was then the Department of Mechanical Engineering and Materials Science. There, he continued teaching materials science including solid state thermodynamics and advanced reaction kinetics. Following his return to WSU, he taught logic design and conducted research in that area until retirement in 2004. Currently, he is Professor Emeritus of the School of Electrical Engineering and Computer Science at WSU.

Index

Printed in the United States
by Baker & Taylor Publisher Services